Workplace ergonomics:

a practical guide

by Dr Céline McKeown and Michael Twiss

Published by IOSH Services Limited

ABOUT THE AUTHORS

Dr Céline McKeown BA (Hons) MSc PhD MErgS AFBPsS MIOSH CPsychol

Céline completed her honours degree in Psychology at University College Dublin and her MSc in Industrial Psychology at the University of Hull. She then joined the Cranfield Institute of Technology's Ergonomics Department, where she lectured in ergonomics to MSc students and presented courses dealing with practical ergonomics to delegates from industrial and commercial organisations. Céline also worked as a consultant on the Department's contracts centring on upper limb disorders and back injuries. She completed her PhD investigating work-related causes of upper limb disorders while at Cranfield.

Céline now works as a consultant, specialising in work-related injuries – particularly upper limb disorders and manual handling injuries. She works in both light and heavy industry, as well as in office-based organisations. Her projects have included workplace assessments – incorporating evaluations of workstation designs, working practices, tools and equipment – and the provision of recommendations for design improvements.

She regularly presents training courses to engineers, planners and architects. She also trains operators, supervisors and managers on reducing the risk of upper limb disorders and manual handling injuries. She is a regular speaker, writer and reviewer on the causes and consequences of work-related injuries and has acted as an expert witness for both claimants and defendants. Céline has assisted in the production of a number of television programmes relating to upper limb disorders.

Michael Twiss BSc (Hons) MErgS MIOSH

Michael has an honours degree in Occupational Health and Safety from the University of Aston. Following graduation, he spent two years in research at the Accident Research Unit of Birmingham University studying how car design influenced the likelihood of occupant survival in vehicle collisions.

Michael spent six years at the University of Loughborough managing a research and consultancy group specialising in the application of ergonomics to health and safety issues.

He now works as a consultant, involved in health and safety management, safety management systems, priority-setting, policy and documentation development and legislation compliance. Michael also deals with workplace design and evaluation, including ergonomics audits and assessments resulting in reports on workable remedies and recommendations. He presents general awareness training as well as more specific courses on safety and hazard management, risk assessment and ergonomics.

Michael has acted as an expert witness on both sides in work-related personal injury cases. He has particular experience of upper limb disorders resulting from office or industrial activities.

Michael is a member of the Ergonomics Society's Health and Safety working group. He has lectured and supervised projects at undergraduate and post-graduate levels and has written a number of academic papers and magazine articles. He is an associate to the Universities of Loughborough, Surrey and Glamorgan.

PUBLISHER'S NOTE

IOSH and IOSH Services Limited assume no responsibility for the contents of this book, in whole or in part, nor for interpretations or concepts advanced by the authors.

All rights reserved. No part of this publication may be reproduced in any material form (including photocopying or storing it in any medium by electronic or photographic means and whether or not transiently or incidentally to some other use of this publication) without written permission of the copyright owners. Applications for the copyright owners' written permission to reproduce any part of this publication should be addressed to the publisher.

Warning: The doing of an unauthorised act in relation to a copyright work may result in both a civil claim for damages and criminal prosecution.

© Copyright Dr Céline McKeown and Michael Twiss 2001

Printed in England by the Lavenham Press Ltd

ISBN 0 9013 5726 X

Crown copyright material is reproduced with the permission of the Controller of Her Majesty's Stationery Office.

CONTENTS

Introduction **7**

The design process **11**

Anthropometrics **39**

Display screen equipment work **57**

Hand tool design and use **77**

Job design and work organisation **91**

Manual handling **105**

Upper limb disorders **135**

Working with external ergonomists **153**

Index **159**

INTRODUCTION

What is ergonomics?

The definitions of ergonomics are many and varied. Some of these definitions, in their attempt to be comprehensive and all embracing, become too complicated to be of any real value despite their apparent accuracy. The succinct title of Etienne Grandjean's seminal 1963 text, 'Fitting the task to the man', is probably comprehensive enough for most purposes and provides a workable definition for this book. The key thing to understand from this definition is that ergonomics is not generally about fitting people to their work. The focus is on ensuring that the work is fitted to the people and is compatible with their abilities and limitations.

It is generally agreed that ergonomics developed as a co-ordinated subject during World War II. The success of the integration which occurred between scientists, engineers, psychologists, physiologists and doctors meant that the multi-disciplinary approach taken continued after hostilities were over. The word 'ergonomics' is derived from the Greek words 'ergon' (work) and 'nomos' (law) and was adopted in 1949 at the foundation of the Ergonomics Society in the United Kingdom.

It is probably worth noting that in the United States of America the word 'ergonomics' is not in common currency – the term 'human factors' is used instead.

The principal function of the ergonomist is to ensure that the tasks undertaken by people are suited to their needs. One of the problems of applying ergonomics principles to any work situation is that it is often seen as exercising common sense. While this may appear to be true, often with the benefit of hindsight, it is frustrating to see just how little consideration has been given to the human component in many work systems.

This book contains many suggestions on how to tackle a variety of health and safety-related issues in the workplace. However, it is not a recipe book and cannot be used to fix all the problems that may be encountered. Its purpose is to make the reader think. Above all, ergonomics is an approach to working systems where the main perspective is that of the humans in those systems. This book will have been a success if it makes the reader understand and adopt the perspective of the human elements.

While health and safety may be the primary focus for many readers, it should be understood that good ergonomics is also good business. Happy, contented and healthy workers contribute to efficient and profitable businesses.

Ergonomics – the process and its scope

Every single person in an organisation needs a measure of awareness in relation to ergonomics. Chief executives, senior managers, designers, architects and procurers, at all levels, should understand the implications of the decisions they make. Many of the decisions made by these groups of individuals have a profound and often catastrophic influence over work systems long before any production takes place.

INTRODUCTION

In reality, there are numerous ergonomics issues arising from the creation of work systems that will need to be addressed before production commences. Once production is underway, its success will only be guaranteed if the middle managers, supervisors and operators who run and use the system are also aware of ergonomics. While the content of this book has implications for everyone within the organisation, it is primarily aimed at people advising on, or having responsibility for, the health and safety function. To that end, the following discussion will concentrate on the role of ergonomics and how it can be incorporated into their work.

The first part of the process should be to identify or predict ergonomics problems. This can be achieved through methods such as observation, user trials, discussion groups and reviews of accidents and ill health and quality problems. The next part will require that these problems are evaluated or assessed to gauge their relative significance. The assessment process can either be undertaken proactively when considering new designs and processes, or reactively by making judgments about situations that already exist. This assessment approach is consistent with the risk assessment methodologies and risk management strategies already used by all competent health and safety professionals today. Following the assessment process it will be necessary to recommend to management appropriate remedies to rectify the ergonomics problems that have been identified or predicted. Clearly, recommendations in themselves are no guarantee of change. It is critical that these recommendations lead to workable remedies.

Following implementation of the remedies it is important that the situation is closely monitored because it is not possible to guarantee on every occasion that the selected solutions were necessarily the best ones available. The choice of solution(s) to be implemented depends to a great extent on the attitude of management towards the remedies, the resources that may be available and the legal duties that are imposed. However, the ultimate success or failure of a new system will often be dependent on monitoring, which may lead to alteration or adjustment before the right solution is found.

The Health and Safety Executive (HSE) is currently encouraging employers to adopt a more participatory approach towards ergonomics. It is very often the case that the workers who actually undertake the task are those with the most intimate knowledge of the strengths and weaknesses of the work system and therefore they are best placed to indicate solutions which can be implemented by the management team. If this participatory approach is to succeed, it is vital that those within the organisation with ergonomics awareness take the trouble to educate and instruct their colleagues so that everyone's awareness is raised.

Increased awareness across the whole of the workforce can have some drawbacks – for example, if different groups come up with their 'own' ideas – and it may fall to the reader of this book to mediate between groups, or to negotiate on their behalf, to arrive at an appropriate solution. The advantages of the participatory approach massively outweigh the disadvantages. In addition, the approach should reduce the organisation's dependence on external consultancy input, although this may be needed occasionally to deal with complex or otherwise unresolvable issues. The chapter entitled 'Working with external ergonomists' gives advice on how to derive the maximum benefit from working with ergonomics consultants.

What should the ergonomics-aware person evaluate?

Greater awareness of ergonomics issues will encourage the evaluation of workstation design, work design, equipment design, work organisation, the environment and the people from a 'user' perspective. Workstation design will need to be based on such things as the requirements of the task, anthropometric data and the standards and guidance applicable to the work. Issues such as the working height will need evaluation and factors such as whether the work is undertaken sitting or standing, whether it is light or heavy and whether visual acuity is important will have to be considered, along with the tools to be used and the demands of the product or process. Consideration will need to be given to reaching distances – whether they involve reaching forwards or backwards, or above, below or to either side of the workstation. In addition, the static or dynamic nature of the work will need close evaluation. The work design will need to be scrutinised, focusing, for example, on whether the work is repetitive or simply monotonous, the cycle times, the effort or force required and whether the work is self- or system-paced. Equipment design may need to take into account aspects such as the tools used, the controls and displays, any racking, stillages or containers that form part of the work system, the use of handling aids and the role of personal protective equipment.

How the work is organised will also be very influential. Areas to consider will include whether the operators work on their own or as part of a team, whether there are opportunities for job rotation, what the work rate is and what the frequency and duration of rest breaks are. There are also likely to be wider but equally important management issues relating to the roles of recruitment and supervision, the identification of training needs and the positive or negative influences of any reward schemes.

The ergonomist will also consider environmental issues such as lighting, noise levels, temperature and vibration – even though these areas may already have received attention from the organisation's safety professionals, occupational hygienists, or occupational health function. Finally, the ergonomist will need to consider the people within the system and focus on such things as their age, gender, size, competence, training, background, habits, medical history and hobbies. These could all be crucially important to their health, safety and welfare.

This book will address all these assessment issues and, hopefully, it will provide the reader with a level of understanding that will allow the introduction of ergonomics principles into work systems with increased competence and confidence.

THE DESIGN PROCESS

Introduction

Those who design workplaces have a significant influence on operator well-being and performance. Unfortunately, many designs are driven solely by the desire to achieve an end result, whether it be the processing and inputting of data from thousands of slips of paper, the production of a chocolate gateau, or the assembly of a fridge. As a result of this approach to design, the operator is often the last element to be considered in the equation, frequently being 'fitted in' towards the end of the process. This short-sighted approach will almost certainly result in mismatches between the final design and the intended user population.

Errors, poor quality, near misses, general dissatisfaction and ill health such as backache and upper limb disorders (ULDs) can often be traced to poor workstation and equipment design. However, having identified the source of such problems, it is usually the case that it would be too difficult, and costly, to change them. The logical conclusion is to develop the right designs from conception, as opposed to after construction.

Ensuring that designs are suitable, usable and safe requires consideration of the individual's characteristics and abilities, and the job requirements. It is important to take a combined approach to design. And, as this chapter is concerned mainly with the ergonomics of design, it is essential that designers refer to relevant regulations and standards before determining any final specifications.

The person

Human operators have certain basic needs which must be met if they are to function efficiently, comfortably and safely at work. Many of these needs are derived from personal characteristics such as body size, and personal qualities such as fitness and strength. These factors should be examined before items of equipment or workstations are designed for use by the operator.

Posture

Operators will commonly adopt a posture which is dictated by the design of their workstation or equipment. Such postures are not necessarily the most comfortable and the least fatiguing. For example, it is not unusual to have control panels in production areas which are set at such heights from the floor that operators have to raise their hands to and above shoulder height when inputting information. They also have to look upwards for extended periods to read the displays. Working in such 'unnatural' postures can lead to pain and sometimes injury, and their significance cannot be overlooked or excluded from the design process. No designer should use the excuse 'we've never had any complaints before' as a means to carry on regardless and ignore the issue. Just because operators are not vocal about concerns, or because there are no recorded injuries or increases in absence levels, the workplace is not necessarily without its problems.

In general, all designs should aim to promote good working postures which do not require high levels of static effort to sustain them. Static effort results from the tensing of muscles for an extended, uninterrupted period of time. The reader can experience this easily by outstretching the arm in front of the body and holding it in that position. This becomes hard work very quickly. Working posture should allow an operation to be completed effectively with the minimum of muscular effort.

Decisions regarding the use of sitting or standing postures should be made early in the design process. The designer should remember that standing requires levels of static muscle work to keep the joints of the knees, feet and hips in a fixed position. Such static muscle work is eliminated when the operator is seated. There are also other benefits:

- the weight is taken off the legs;
- the chances of adopting an irregular or unnatural body posture are reduced;
- the operator has greater stability;
- jobs requiring fine, precise or manipulative movements may be carried out more easily, particularly if the arms are supported;
- energy consumption is reduced; and
- demands on the cardiovascular system are reduced.

Despite these benefits, there are some disadvantages to sitting at work – in particular possible increases in backache. However, it must be said that backache is often directly related to the use of a poorly designed chair and/or the inappropriate use of a chair. Both these areas are dealt with later in this chapter.

If a workstation is specifically designed to be used by a seated operator, it should not, at a later date, be used from the standing position unless it has been altered accordingly. More often than not, the height of a seated workstation will be lower than that for a standing workstation, therefore standing operators will spend their time leaning forwards as they reach down towards the worksurface – a common cause of backache. However, a workstation designed for the standing operator can always be used by the seated operator, providing the work allows it and the workstation design is compatible with the chair use. One of the most important considerations in this situation is a height-adjustable footring or footrest as the operator will be sitting at an elevated level in relation to the floor.

There are several basic rules which, if incorporated into a design, will ensure that an operator can work comfortably and effectively whether sitting or standing. These are:

- the operator should be able to adopt an upright and forward-facing posture;
- if standing, the body weight should be borne by both feet equally;
- the posture should be balanced so that additional muscle activity is not required to support or stabilise the body as a whole, or individual limbs, such as would result from leaning forwards;
- the head should remain reasonably upright or slightly inclined to the front;
- the limbs, trunk and head should be positioned during the work activities so that the joints are not forced to go beyond their mid-point in terms of their range of movement;
- the hands should not have to pass above elbow height on a regular basis or for extended periods of time;

- the largest (appropriate) muscle groups should be used to apply necessary forces in a direction which is compatible with their structure, eg using the leg muscles to operate a foot control in place of the use of a hand control;
- the workload should be shared out equally/evenly across the body as a whole and between individual limbs; and
- the work should be capable of being carried out when the operator is in varying postures (and this should be achieved without a change in the quality of performance or comfort of the operator).

To ensure that such general principles can be applied in the workplace, it is important that designs suit the size and shape of the user population. Only then will operators be able to adopt and maintain 'natural' comfortable postures.

Anthropometrics

Anthropometrics is concerned with body size (see the chapter entitled 'Anthropometrics'). The main point to remember during any design process is that designs should not cater for the 'average' person due to the number of potential users who will be put at a disadvantage when using the workstation or equipment. Consideration of body size and related influential factors, such as gender, will enable the designer to calculate appropriate dimensions when determining acceptable reaching distances, overhead clearance, working heights and so on. The word 'calculate' is used deliberately because the decision regarding dimensions is not, or should not be, a hit and miss or trial and error affair with the designer assuming that because a dimension 'fits' them, it will suit everybody – in the same way that no-one would assume that just because a jacket fits them comfortably it will fit everybody else equally well.

Using the concept of anthropometrics the designer can plot dimensions which are important to the user of a workstation or piece of equipment.

Access

Before an operator can even take up position alongside their workstation or equipment, they must be able to come within range of it easily, without any deviations from their upright posture and without having to extend their reach. To ensure ease of access, the design should provide enough space for the largest possible operator who may work in that area. This will necessitate consideration of factors such as operator stature, shoulder and hip width, and dimensions relating to hand and arm size. Additional allowances will have to be made for the movement of tools and other equipment or components by the operator, as well as for the storage of other items in the area. Individual differences will have to be taken into account where appropriate or necessary. For example, access requirements for a person in a wheelchair will be different to those for someone who walks to and from an area.

Designers should also acknowledge the fact that the equipment or workstation will probably not function perfectly 100 per cent of the time it is in use. They need to take account of the fact that there will probably be breakdowns at some stage in the workstation's or equipment's lifetime. Therefore, when working on the concept of access, consideration should be given to the requirements of maintenance personnel. Allowances should be made for the largest parts of the equipment or workstation that may have to be moved during repair or rework, as well as any tools that may have to be used. Of course, the operator may also be expected to respond to equipment breakdowns. For example, it is common practice in the food industry to 'bulk-off' when a line malfunctions. Food cannot, for obvious reasons, be left lying around on conveyors until the line is up and running again. It has to be removed quickly and placed in chiller areas (ie the bulking-off process). However, more often than not, access along the conveyor will not have been built in at points where operators are required to remove the food. As a result, they will spend the bulking-off period reaching across large side casings on the conveyor, or over ingredient hoppers, or across conveyors of such a width that their upper body is almost perpendicular to their lower body as they attempt to stretch to the far side. Allowing for such events should be a fundamental part of the design process.

Clearance

Once in the work position, the operator should have clearance in all directions. Overhead or vertical clearance will ensure that the operator does not have to stoop. When determining this factor the designer should cater for the tallest operator who could be expected to work in the area. Having reviewed the dimensions relating to stature, consideration should be given to the use of headgear and footwear as both will have a significant impact on the level of overhead clearance.

Side-to-side and forwards-backwards clearance is also important. Again, the designer should cater for the largest operators who could be expected to work in the area, making allowances for some movement during the activities. Hip and shoulder width and trunk dimensions are important. In addition, the use of any other equipment will have to be accounted for.

Foot clearance is commonly overlooked in the design of workplaces. If a recess is not provided at floor level so that the foot can move beneath a workstation frame, the operator will be held back from the leading edge of the worksurface. The operator will overcome this deficiency by leaning or reaching forwards. Further allowances should be made if a pedal is in operation.

If it is intended that operators will sit down when working, it is important that they are provided with adequate clearance for their knees. When determining the required clearance, consideration should also be given to hip width and the width of the seat being used. The provided clearance should cater to the longest thigh. As it is unlikely that the operator will adopt a fixed position when working, an allowance for leg movement should also be incorporated into the design which will require a review of overall leg and foot lengths. Vertical clearance between the seat's surface to underneath the worksurface should not be ignored. An allowance should be made for leg crossing or the use of pedals which may result in the raising of the knee. Again, the largest thigh should be taken as the benchmark. If undersurface clearance is not provided, operators will be held back from the worksurface and will have to lean and reach forwards when completing an operation. Alternatively, operators faced with a lack of undersurface clearance may choose to sit sideways to the workstation and twist their upper body in the required direction. Lack of knee-room is usually the major shortfall in a workstation which was designed for the standing operator but which is used at a future date by a seated operator.

Reach

When determining reaching distances, the aim should be to allow the operator to grasp an object without excessive body movement or energy expenditure. This is particularly important in the case of seated operators who cannot move towards an object or work area. When examining reaching distances, the 'zone of convenient reach' and the 'normal working area' should be considered.

Zone of convenient reach

The zone of convenient reach is sometimes referred to as the secondary work envelope or maximum reaching distance. Areas which can be reached easily when the arms are fully extended sideways, forwards and upwards fall within the zone of convenient reach. The illustration below shows two intersecting curves. The radius of each curve is the length of the arm (a) and the centre of each curve is separated by the biacromial width of the shoulders (b). The sagittal plane (sp), transverse plane (tp) and coronal plane (cp) are also shown.

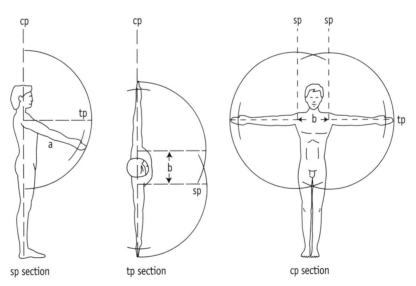

Zones of convenient reach

Reproduced from 'Bodyspace – anthropometry, ergonomics and the design of work' (Pheasant 1996)

A work area or an object required during an operation should not be located beyond the zone of convenient reach. However, working with the arms at their maximum range of extension repeatedly or for prolonged periods is fatiguing and can be damaging for the limbs. Equally, operators should not be required to work with their hands and arms towards the rear of the body, nor to have to complete repetitive or sustained operations with their hands above shoulder height. While it is acceptable to undertake occasional activity within the zone of convenient reach, sustained work needs to be undertaken much closer to the body.

THE DESIGN PROCESS

Normal working area

The normal working area is sometimes referred to as the primary work envelope or optimum reaching distance. It is smaller than the zone of convenient reach and is defined as the area that can be swept comfortably by the forearm when there is a 90° angle at the elbow and the upper arm is hanging down naturally. Objects required frequently or actions that are undertaken for sustained periods should be kept within the normal working area. The zone of convenient reach and the normal working area are shown below.

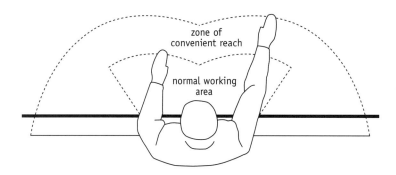

The relationship between the zone of convenient reach and the normal working area

When setting dimensions for optimum reaching distances, primary consideration should be given to the shortest arm length (includes upper arm, forearm, hand and fingers). If dealing with extended arm length (to the end of the fingertips), the designer should recognise that this dimension is reduced if the operator grips an object, due to the fact that the fingers are closing into a fist and enveloping the object.

Dimensions for British adults aged from 19 to 65 years are included in the 'Anthropometrics' chapter. Use of the data will allow calculations of the zone of convenient reach and normal working area.

Reaching distances can also be applied to feet and legs, particularly if a pedal is being operated from the seated position. The designer should cater for the shortest leg length as well as the longest. Adjustability in pedal location will be of great benefit.

Personal space

When designing several workstations with the intention of locating them in close proximity, the designer should recognise the effects of 'territoriality' or personal space. Invasion of an individual's personal space may undermine the acceptability of a design. 'Personal space' is defined as a portable, invisible boundary which surrounds an individual and into which entry is restricted. This boundary is divided into several zones, each of which is determined by levels of social interaction. For example, people on an intimate level will be allowed closer than those on a purely friendly level, who in turn will be allowed closer than those on a wider social level, and so on. Should someone invade a zone within another's personal space which is not appropriate to their relationship, stress or anxiety can be experienced. For example, most people will accept accidental elbow contact with family members when eating a meal at home but will be uncomfortable when similar contact occurs with a stranger sitting in close proximity in a restaurant.

Personal space is not an evenly spread zone which surrounds the body on all sides. For example, personal space zones can be invaded to a greater extent from the sides than from the front. Consider a passenger standing on a crowded bus – a stranger will be more acceptable when standing at the side than if in front and facing them. The latter is deemed to be much more threatening. In addition, the appraisal of acceptable 'uninvaded' space changes if the space above alters, ie if a ceiling is low, people will increase their personal space outwards away from the body.

Handedness

Operators who use a workstation or piece of equipment may not be right-handed. Consideration should be given to producing designs, if practicable, which can be used equally well by either left- or right-handed operators. Designs which force an operator to use their non-dominant hand may cause problems.

If a design is intended to be used by either right- or left-handed people, its layout should facilitate such usage.

Disabilities

The Disability Discrimination Act 1995 defines disability as: "A physical or mental impairment which has a substantial and long term adverse effect on a person's ability to carry out normal day-to-day activities." In this context, 'impairment' includes:
- physical impairments affecting the senses, such as sight and hearing; and
- mental impairments including learning disabilities and mental illness.

Disabilities arise from 'impairments' which themselves are considered to result from the interference of normal functioning or structure. Such interferences can derive from illness, injury or genetics. The introduction of the Disability Discrimination Act has significant implications for employers, service providers and others. For example, in the workplace a designer may have to take into account the specific needs of wheelchair users, or operators who have lost limbs or who have impaired vision or hearing. There are other levels of disability (eg backache or arthritic conditions) which may not be covered by the legislation but which may also need to be considered.

Specific data are available to assist in designing for wheelchair users (see, for example, Pheasant 1996). As is the case when designing for able-bodied operators, it is much more efficient and cost-effective to design for disabilities in the early stages rather than to try to alter an environment post-construction.

IN SUMMARY

Posture
- Postures should not require high levels of static effort to sustain them.
- Variations in working posture should be permitted but not at the expense of operator comfort or performance.

- Designers should consider the use of either sitting or standing postures early in their decision-making.
- Sitting down while working reduces the degree of static work required of the body.
- The operator should be able to adopt an upright and forward-facing posture.
- If standing, the body weight should be borne by both feet equally.
- The posture should be balanced so that additional muscle activity is not required to support or stabilise the body as a whole, or individual limbs, such as would result from leaning forwards.
- The head should remain reasonably upright or slightly inclined to the front.
- The limbs, trunk and head should be positioned so that the joints are not forced to go beyond their mid-point in terms of the range of movement.
- The hands should not have to pass above elbow height on a regular basis or for extended periods of time.
- The largest (appropriate) muscle groups should be used to apply necessary forces in a direction which is compatible with their structure.

Anthropometrics

- Designers should consider the body size of the user population.
- Designs should not be based on the 'average' person.

Access

- Operators must be able to come within range of their workstation or equipment easily.
- Access dimensions should cater to the largest operator who may need to work in an area.
- Allowances should be made for the use of any tools or other equipment in the area, as well as space requirements for storage of items.
- Allowances should be made for maintenance operations and the personnel who will carry out repair work, the tools they may use and the largest parts of the machine or workstation that may have to be moved.
- Consideration should be given to the work to be carried out by operators following any system failure – access should be provided at relevant points in the work area.

Clearance

- Vertical or overhead clearance should cater for the tallest operator who will work in an area – allowances for any headgear or footwear should be included.
- Side-to-side and forwards-backwards clearance should also cater for the largest operator who may work in an area.
- Foot clearance is required at floor level so that operators can move closer to worksurfaces or equipment (if necessary, adjustments should be made for foot pedals).
- Seated operators should be provided with knee clearance that caters to the largest hips, the width of the seat and the longest thigh, and permits forwards or upwards leg movement.

Reach

- Frequent or sustained work should remain within the operator's optimum reaching distance (defined

as the area that can be swept easily by the forearm when there is a 90° angle at the elbow and the upper arm is hanging naturally at the side of the trunk).
- Designs should cater to the shortest arm length and allow for the fact that the dimensions for extended arm length are reduced if the operator grips an object.
- The reaching distances to items such as foot pedals should be matched against user leg lengths.

Personal space
- Personal space should be taken into account when designing several workstations or pieces of equipment to be located in close proximity for use by several operators.

Handedness
- Consideration should be given to the use of designs by both left- and right-handed operators.
- The ability to use two hands simultaneously or alternatively should be incorporated into the design.

Disabilities
- Disabilities should be considered and catered for during the design process.

The task

A task may require an operator to work in a particular manner, to use certain tools, to operate controls and to read information. These requirements will dictate elements such as the height of the worksurface, reaching and viewing distances and the shape and location of controls. Therefore, before designing any workstation or piece of equipment, the designer should have a clear understanding of the task demands. To ensure that a clear profile of all these requirements is developed, a detailed task analysis should be carried out.

Task analysis

A task analysis allows the job to be broken down into parts, enabling the identification of all the sub-routines involved in an activity. This allows the designer to map out all movements, postures and forces required during the activity as well as the information required by the operator. It is only by completing such a detailed and accurate task analysis that the designer can develop a suitable specification which matches the job requirements.

During the task analysis information can be gathered using a range of sources, including:
- **Examination of current documentation** such as operational manuals, training manuals or previous reports relating to the work. This will usually provide detailed accounts of how work should be completed. However, operators do not always work in a manner which mirrors instructions in a manual and examination of documentation should be supplemented with other forms of analysis.
- **Observational methods** – the observer should watch the operation and document the activities

using a checklist, flow chart or descriptive format. Timing of the operation is necessary but only as a means of determining priority or duration of tasks. This is not an exercise to try to find ways of shaving a few seconds off the processing time.
- **Recording with cameras and videos** – this has the advantage that the work can be studied repeatedly, if necessary, without having to spend long periods at the work site or without asking the operator to repeat the same portion of the operation.

When completing a task analysis, consideration should be given to any unseen activities which may occur irregularly, for example, maintenance or other work completed on unobserved shifts. These activities should be reviewed before any final conclusions are drawn.

Working height

Working height is defined as the level at which the hands work. This is not always the same height as the worksurface. As a general rule the operator should be able to complete their activities with their hands at, or slightly below, elbow height whether they are seated or standing. Therefore the elbow height of the operator should be used as a guideline for determining worksurface height.

If the user population presents the designer with a range of varying elbow heights, the taller operators should be catered for and platforms or duckboards should be provided for shorter operators.

Some designers may raise objections about having to provide a range of platforms of different heights. Given that operators can generally work in the range that falls at or slightly below elbow height, if the most appropriately dimensioned platform is used, operators of different statures using this single platform should find that it allows them to work somewhere within that optimal range. It may not therefore be necessary to provide platforms of varying heights.

By providing platforms, further problems, such as tripping and hampering the transport of items around the area, are introduced. In addition, other practical issues, such as the cleaning of these platforms, their manual movement and the additional workspace required to position them, will have to be taken into account.

The need/desirability of presenting a worksurface at elbow height is not the only consideration when determining working heights. For example:
- If operators are involved in an assembly operation they may locate the components for assembly on their worksurface. If the worksurface has been set at elbow height, the addition of the parts to the worksurface increases the working height above elbow height. This working height is increased further if the parts are placed in a jig and if the operator uses a tool, such as a drill, during the assembly process. It is quite common to walk around an assembly area and find that most operators are working with their shoulders raised and their elbows sticking out to the sides. Such postures are indicative of a working height which is too high.
- The force requirements of a task will also influence the appropriateness of worksurface height. If operators have to apply a large degree of manual force when assembling a part, or if they are using a tool such as a hammer, they may find that a worksurface set 10–15cm below elbow height is more

acceptable. However, the worksurface should not be set so low that the operator ends up stooping. If the heavy work is only one part of an overall operation it may be necessary to provide the operator with two surfaces providing different working heights.

- If the operator is carrying out fine or precise work they may require the worksurface to be raised above elbow height. In such cases, a worksurface set 5–10cm above elbow height may be of benefit. Tasks requiring high levels of visual acuity may need to be performed slightly below shoulder height so as to avoid the need for the operator to lean forwards and downwards towards the workpiece. The use of padded supports for the arms, hands or wrists should be provided in such a situation to reduce the loading on the shoulder muscles and to steady the hands. The design of these supports should be such that they do not hinder the operator. A level of adjustment in the pads may be advisable. In addition to raising the worksurface for fine work, consideration should be given to tilting the surface towards the operator. This will further minimise the need to lean forwards towards the work area. It will also reduce the potential for seated operators to raise their arms and bend their wrists when working.

Visual requirements

The visual demands of a task will have a direct bearing on head and neck postures. If operators are required to read or inspect an item during the course of their work it is important that the object of attention is presented in an appropriate orientation. If this is not the case the operator will have to alter their posture to complete the task.

The eye usually adopts a downward cast of about 15° below the horizontal when resting and when the operator is seated. When standing, the downward gaze is usually about 10° below horizontal. These ranges represent the normal line of sight. However, the eye has a total range of movement of about 60° below the horizontal. The most comfortable viewing is achieved within the band ranging from horizontal to 30° below the horizontal. This range has been referred to as the preferred zone for visual displays but it has been recognised that the zone covering up to 15° below the horizontal is the best area for attention, scanning, viewing objects at a distance and viewing details and colours.

Non-priority visual tasks can be performed in the less favourable visual zones, ie outside the 30° range, and can involve occasional but small head and eye movement. Very low priority tasks, carried out on an irregular basis, can call for the movement of the trunk also. The priority of the task determines the location of the source of information and the postures to be adopted by the eyes, head, neck and trunk. Therefore, given its importance, the priority of the task should be quantified early in the design process.

A viewing distance of 350–500 mm is considered acceptable, however, this range should be reduced if the work is detailed, and extended if the object is large.

Efforts should be made to ensure that obstacles do not block free viewing by the operator. Designers should take account of operator stature, both seated and standing, when determining the effect of obstacles, as objects in the visual field may become obstacles as an operator's eye height is reduced, ie a short operator may not be able to see over other objects in the visual field to focus on the target object.

As a general rule, high priority, frequent or sustained work, or work requiring high speed scanning or high levels of accuracy, should be positioned so that it can be read/attended to when the head is upright (or slightly inclined), facing forwards and in a relaxed position.

Where there is a need to look at very detailed work, magnification should be considered.

Controls

If the task requires operators to use controls during their activities, these controls should be given as much consideration as any other workstation or equipment element. Controls are the means with which operators interact with the system, therefore if they are to respond appropriately to work demands they need to be able to use the controls efficiently and accurately. And, given the critical nature of control use in some situations, it is obviously essential that operators are able to differentiate between controls, adjust them to the appropriate degree and operate them correctly even (or especially) in emergency situations.

As in the case of designing a workstation or piece of equipment, a task analysis is required to determine the final form a control should take. The task analysis will allow knowledge of significant aspects such as the forces used, the speed and accuracy required, the pattern and priority of usage, and the frequency and duration of use. These will guide both the design and location of the control. The following important aspects should be considered during control design:

Force

If limited or low levels of force are required, such as in the case of an on-off switch, then a toggle switch, push button or key-lock would be acceptable.

If greater degrees of force are required, the design should allow for the force to be applied easily. In such situations reliance on the fingers alone to complete the action would not be appropriate, therefore the control should take a form which allows for the application of force using the whole hand, such as in the case of hand-push buttons, levers, joysticks, cranks, foot pedals and handwheels.

The duration of the effort should also influence the final form that the control takes, and consideration should be given to the 'weakest' operator, taking into account age, gender, fitness and training.

Feedback

The guidance of a machine is likely to be more accurate if it moves in a direction and at a speed that corresponds to the movement of the control. If the operator experiences high levels of resistance when using a control, or if there is little obvious movement or sensation of movement in the control, the operator may not be sure that the desired movement has been achieved. However, if this is corrected by making a control more sensitive it should not then be set at a level which permits accidental activation.

Layout

High priority controls should be located in the primary work envelope directly in front of the operator. If they are intended to be used by both upper or lower limbs simultaneously they should be located such that they are accessed easily and equally by both limbs. If the intention is that both left- and right-handed operators will use the controls, they should be located centrally. This will also allow for use by either hand by any single operator.

Function

Controls with a similar function should be grouped together, in order of priority. However, they should not form more than a row of three and a column of three. Controls in a group should not interfere with each other during use.

If controls are used in sequence during an operation, their layout should match the sequence and their usage should be accomplished smoothly. A standardised left-to-right, top-to-bottom order is the most acceptable. If the sequence is unchanging, consideration should be given to interlocking the controls which will make their operation more efficient and will control the potential for errors. If particular controls are to be left out of a sequence on occasion, the potential for erroneous inclusion in the sequence can be avoided by coding of the control (see below).

Expectations or stereotypes

Expectations will influence the acceptability of a design and the error rate associated with its use. Operators may expect or assume that a control will be in a particular location. As a consequence, any 'blind' search and use may lead to errors. Cultural stereotypes will 'train' an operator to expect certain responses from a control, for example, flicking the switch upwards typically turns the system off in Europe but turns it on in America.

Designing contrary to stereotypes and expectations can lead to critical situations, particularly during emergencies when the operator is subject to stress and prone to reverting to the stereotype. Such designs may also lead to longer training times and reduction in speed and accuracy during early usage – and, even with experience, reaction times will be slower.

Compatibility

Operators will expect controls to move in a direction which is compatible with the layout of a display or system. They will also expect controls on similar machines to operate in a similar manner. Compatibility is particularly important in situations where:

- the operations are complex;
- training has not been extensive;
- sequences of use are not always the same;
- the operator is responsible for monitoring and interacting with a range of machines; and
- where any mistakes are costly or dangerous.

Coding

Controls can be coded by adding features which allow them to be distinguished more easily. For example:
- **Location** – as already discussed, operators have expectations of where they think a control should be located.

- **Shape** – it has been shown that in conditions of stress, an operator can still distinguish up to 12 different controls by their shape. This is particularly the case when the shape of the control offers an indication of its function.
- **Size** – it has been shown that operators can distinguish up to three different sizes of control in moments of stress, suggesting the use of small, medium and large controls.
- **Colour** – different colours may be used to distinguish controls. However, such a system has disadvantages: it will only work efficiently in well-illuminated environments; it takes longer for the operator to respond as the meaning of the colour has to be interpreted before action can be taken; and cultural stereotypes may interfere with appropriate reactions (eg red generally represents 'danger' or 'stop' in the UK while in China it is associated with happiness). Thought should also be given to the effects of colour blindness in operators.
- **Labelling** – written labels can describe the function of a control. To be effective the label should be capable of being read easily when standing or sitting in the position where a control will be used. It should be read from left to right (vertical instructions take longer to read). Labelling has its disadvantages: labels may not survive in a legible form in a very dirty or wet environment (embossed labels should be used); the success of the instruction will be determined by the literacy level of the reader; and (as with colour coding) the information has to be interpreted before action can be taken, hence increasing reaction time.
- **Relation to displays** – controls and their related displays should be positioned so that both can be viewed simultaneously without the need for adopting an irregular posture. To facilitate this controls should be close to the display or at least grouped in a similar pattern to that of the display. Layouts of groups of controls and displays should be compatible, for example, the bottom left control referring to the bottom left display. Relationships can be highlighted by borders around displays and their related controls, or by placing both on a coloured panel.

Displays

The important requirements of any display are that the desired information can be read and understood easily, rapidly and accurately. Therefore, to produce the appropriate design, thought should be given to the type of information being presented, the levels of interpretation and response required and whether the information has to be memorised. The location and layout of displays should be given similar consideration to that for controls (discussed above). Additional points for deliberation include:

- **Dials** can carry a wide range of information but are not considered appropriate for providing warnings or complex information.
- **Lights** can convey information as a stand-alone light, as the illumination for a button or as part of colour coding. They are useful for drawing attention to urgent information or warnings, particularly when flashing.
- **Digital displays** are useful in situations where information changes slowly, numerical data are presented, high levels of accuracy are required and the duration of reading bouts is short. Digital

displays have disadvantages, particularly when information is changing rapidly and when the direction of change or ranges of movement are important.
- **Annunciators** (devices which give a visual indication as to which of a number of electric circuits has operated, such as an indicator that identifies exactly which area of a burglar alarm has been triggered) can be useful to confirm that a particular control has been activated, or to provide a signal that action is required.
- **Auditory displays** are commonly used as a means of providing warnings. Their frequency and intensity should be compatible with the environment. Alternatives include systems which are based on tactile (touch), kinaesthetic (movement) or olfactory (smell) information, which are particularly effective if deaf or partially deaf operators are working in the area.

Labels

The most effective label is one which carries information which is short and to the point while still managing to convey a full message. If a message is too long it may not be read. If a message is written in positive terms it is more likely to be acted on than one written in either passive or negative terms. For example: "Pushing the red button will stop the conveyor" (positive); "The conveyor can be stopped by pushing the red button" (passive); and "Pushing the green button will not stop the conveyor" (negative).

The most difficult label to understand is the negative one, as the reader has to understand first what they are *not* supposed to do before they can work out what they *are* supposed to do. An everyday example of this is a road sign which states that parking is not allowed between certain hours. This takes longer to comprehend than a sign which outlines when parking *is* allowed.

A label should be capable of being seen and identified easily even in the poorest working conditions. The reading of the message should be facilitated by using appropriate letter/number size, typographic style, layout and spacing. The use of jargon or a 'bureaucratic' style should be avoided. Simple words and sentences should be used. If abbreviations have to be included they should be readily understood by all operators. Etched and embossed labels will probably last longer than painted or printed ones.

Symbols

In some cases, a symbol may be more appropriate than a written message as it will carry information much more succinctly. In addition, symbols should be more readily understood by operators who do not have a thorough understanding of the relevant language or who do not possess suitable levels of literacy.

One of the disadvantages of symbols is that, although they may draw an operator's attention to a problem situation or provide advance warning about states of functioning, they do not always convey what action is required once the particular situation has been recognised. In addition, symbols are completely redundant if the reader has not received any appropriate training relating to their meaning and the necessary actions required. In this context, there may be some advantages to using pictograms where the appropriate action is demonstrated as part of the graphic.

To ensure that symbol use is successful, the design should use widely agreed and accepted standards such as those produced by the International Standards Organisation. An example of an internationally accepted symbol is that used in a car to indicate which control operates the fog lights.

IN SUMMARY

Task analysis
- Task requirements must be fully understood prior to the design process. They should be identified via an in-depth and accurate task analysis.

Working height
- Working height may be different to worksurface height and is defined as the height at which the hands work.
- When setting surface heights the work should generally be presented to the operator at or slightly below elbow height unless the task requirements demand a different level.
- Adjustments should be made to surface height to allow for the addition of parts, jigs and tools to the worksurface.
- Consideration should be given to other task requirements – such as the forces required and the precision of the work – which may also dictate an alteration in worksurface height.

Visual requirements
- The most comfortable viewing is achieved within the band that ranges 30° below the horizontal eye level.
- Tasks which require attention, scanning, viewing at a distance, in detail and of colours, should be located in the zone which covers up to 15° below the horizontal.
- Non- or low-priority tasks can be located outside the 30° from horizontal range and can involve head, neck and trunk movements.
- High priority, frequent, sustained, high speed or accurate work should be carried out when the head is upright, facing forwards and in a relaxed position.

Controls
- The design and layout of controls should be based on task requirements.
- The force and duration of force will influence the shape of the control.
- If the movement of the control corresponds to that of the machine its operation will be more accurate.
- High priority controls should be directly in front of the operator and located for use by the appropriate hand(s).
- Controls with a similar function or sequence should be grouped together but with no more than three per row and three per column.
- Stereotypes will influence acceptability of a control design and levels of errors during its use.

- There should be compatibility between controls on similar machines and between controls and related displays.
- Coding can be used to distinguish between controls.

Displays

- Displays should be located in a similar manner to controls.
- The type of display should match the requirements of the task and how the information has to be used.

Labels

- Labels should carry short, pertinent information which conveys the full message.
- Information should be in a positive form.
- The label should be constructed to remain legible in the working conditions.
- The style and layout of language should facilitate easy reading of the message.

Symbols

- Operators should receive training in the interpretation of symbols and the actions required following identification of the given situation.

Additional design considerations

Future use

Many workstations or pieces of equipment are not being used as originally intended. This is usually as a result of an alteration in product design or processing procedures. The consequence of this change in use is that many operators are forced to work in highly irregular postures, or apply higher levels of force than was originally planned, simply because the designer had only thought in 'here-and-now' terms and neglected to consider the possible future uses of the design.

If designers are able to build a degree of flexibility into their specifications, the system will be capable of manipulation at a later date without compromising the safety and well-being of operators.

Adjustability

The best method of ensuring that the majority of the working population can use a design without undermining health and safety is to provide it with an adjustment facility. The adjustment feature does not have to be complex or sophisticated to work, nor have to operate through a gas-lift mechanism, crank-handle or electrically powered control panel. Workstations can be fitted with adjustable legs or feet which can be unscrewed to extend their length. This is particularly effective if the same operator will use a workstation for a period of time and it does not have to be altered frequently for other operators. Some workstations can be made from a frame which can be unbolted and adjusted, then bolted together again.

The greater the degree of adjustment and flexibility offered to operators, the more likely they are to be comfortable when they work. The easier an adjustment is, the more likely it will be made by the operator. The appropriate tools should be available to adjust a workstation (eg the right size of spanner).

Sharp edges

Workstations and equipment commonly do not have rounded edges. Operators often lean on these edges and, apart from the obvious outcomes such as bruising or abrasions, compression of the skin will reduce the blood flow into the hands and can contribute to the development of ULDs over time. To minimise the likelihood that cumulative damage to the limbs will result, contact edges should be rounded or padded.

Cold surfaces

Working in, or coming into contact with, reduced temperatures can contribute to the development of ULDs. Therefore, if operators use worksurfaces which are cold to the touch, such as stainless steel inspection and packing lines in the food industry, they should be discouraged from leaning on them. If operators cannot be prevented from coming in contact with these surfaces, the surfaces should be padded.

Seating

Generally, seats in industry appear to be viewed simply as perches on which to rest the buttocks. It is quite common to find operators using chairs which are in such a poor state of repair that they are taped together, or have had packing materials stapled onto them to make them more comfortable. If more effort is put into the design and selection of seating, operators will be less likely to suffer from backache and more likely to remain comfortable throughout the day.

If an operator is required to sit down when working, and to use equipment in the most efficient way possible, it is essential that the chair offers the following features:

- it is adjustable for height;
- the backrest has independent height adjustment;
- the backrest can be altered for inclination;
- the backrest is adequately shaped and includes a distinct lumbar support;
- the seat can accommodate the largest hips;
- any armrests are padded and, preferably, adjustable for height;
- adjustments are easy to reach and use from the seated position;
- it is (ideally) capable of being moved towards or away from the worksurface by the operator rather than being fixed in one location;
- if it is raised to a high point, an adjustable footrest or footring is provided; and
- any pedals are capable of being adjusted for height in relation to the operator's feet.

The workstation design should be compatible with the use of the seat.

Safety

Although the emphasis has been on designing to suit the individual, consideration should also be given to safety issues. When determining appropriate workstation or equipment dimensions, an additional allowance has to be made in some situations to ensure that the operator is at a safe distance from hazards. Such hazards may include parts of a machine which saw, press, spray, weld, cut or eject parts.

When determining safe distances, consideration should be given to whole or part body movement during reaching and leaning, the insertion of a limb through an aperture, or the squeeze point where part of the body is pushed through an access point (eg pushing the leg through a gap between two pieces of equipment).

It is essential that all relevant national and international standards are studied carefully during this phase of the design process.

IN SUMMARY

Future use
- A degree of flexibility should be built into a design to allow for future changes of production which may necessitate different use of the system.

Adjustability
- A good range of simple-to-use adjustments will make it more likely that the operator will be comfortable.

Sharp edges
- Operators should be prevented from leaning on sharp edges of equipment or the edges should be padded or rounded.

Cold surfaces
- Operators should be protected from cold surfaces by either preventing them from leaning on them or by padding them.

Seating
- The chair should be adjustable for height.
- The backrest should have independent height adjustment.
- The backrest should be adjustable for inclination.
- The backrest should be adequately shaped and include a distinct lumbar support.
- The seat should accommodate the largest hips.
- If armrests are supplied they should be padded and, preferably, height-adjustable.
- Adjustments should be easy to reach and use from the seated position.
- The chair should be capable of being moved to and from the worksurface (rather than being fixed).

- If the chair is raised to a high point, an adjustable footrest or footring should be provided.
- Any pedals should be capable of being adjusted for height in relation to the operator's feet.
- The workstation design should be compatible with the use of the seat.

Safety

- Safety considerations should be incorporated into the design process.

Selection and use

Having developed the workstation or equipment design to the point where it is ready for construction, the designer should consider a mock-up and subsequent trial before committing to full introduction of the change to the working environment. This also applies to purchasing decisions regarding a design provided by a supplier. Decisions should only be made following the design's use in a near-normal situation.

Mock-ups

In the first instance a chipboard mock-up should be assembled to test dimensions and layouts. A range of operators of varying sizes should be asked to try the mock-up to determine whether it is suitable for use. It is recommended that a level of adjustment is built into the mock-up at this stage so that operators can try out various settings. This will assist in finally deciding on any single setting if the workstation is to be non-adjustable. If adjustments are to be built into the mock-up, they should be capable of being made quickly so that the operator is able to make a sound comparison of settings.

At the end of this stage the designer will have an indication of maximum and minimum acceptable working limits and areas of convenient reach and comfort which suit the likely range of operators.

Trials

Having established the acceptable working limits and layouts during the mock-up phase, prototypes can be developed and used in a full trial phase. During the trials as many operators as possible should use the design under near-normal working conditions. The operators should be representative of the normal user population in terms of skill, age, size and background. However, allowance for 'individual differences' will have to be made. Such individual differences can result from personality, attitude and level of alertness.

During the trials, operators should alter one dimension at a time (eg working height) between minimum and maximum acceptable levels, making comparisons between each. All comments regarding the design should be recorded in full. Informal verbal feedback which has not been recorded is not a sufficient basis on which to change a design. The designer, or trial manager, also needs to act as an 'action interpreter' during the trials, as operators may carry out movements or change settings without providing feedback or without realising that they have actually done something. All this information should be collected and evaluated.

Once operators have worked with the prototype and provided feedback on acceptable and unacceptable working limits, a degree of overlap should be found between the preferred settings. This will indicate the level at which a dimension can be set. If such an overlap is not apparent it may be necessary to redesign the workstation or equipment or to provide a level of adjustment in the final design.

If a workstation or piece of equipment is provided by an outside supplier it should be placed in the working environment and used 'normally' for several weeks by a range of operators. All feedback should be recorded and analysed at the end of the trial phase.

Layout

Having developed a workstation design, consideration should be given to the layout of its surface. Items are not put on a surface in a random position. Their location should be determined by their use. It is recommended that a distinction is made between primary and secondary items. Primary items are those that are used frequently, and secondary items are those that are used less frequently. Primary items are placed within the primary movement envelope of the operator, ie the area that falls directly in front within their optimum reaching distance. Secondary items are located in the secondary movement envelope which falls outside the primary but does not exceed the maximum reach of the operator. Determining the primary and secondary layout of items will apply to books, files, telephones, tools, waste bins, components and controls. It may also be beneficial to consider laying out work items to suit sub-tasks as well as the overall task.

Training

In order to achieve the most from any new workstation or equipment it is essential that operators are given full instruction on use. If such instruction is not given, operators will develop their own way of using the items which may not be the most suitable or safest method. In addition, there is the possibility that without training on how to use the workstation or equipment, operators may refuse to use them.

Supervision

Having trained the operators in the use of a workstation or equipment it is important that proper use is maintained. This can be achieved through adequate supervision. Supervisors should encourage the continued and appropriate use of any new items and should discourage any misuse or neglect of them. Feedback from supervisors to designers is an important element in continuing good ergonomics design.

Maintenance and housekeeping

New designs introduced into the workplace should be maintained satisfactorily. If this does not happen, the item's condition may deteriorate and eventually undermine its effective use. Operators, in turn, should ensure that poor housekeeping does not result in the new design being more difficult to use. Again, this is an issue for supervisors to deal with.

Environmental conditions

If maximum effective use of a design is to be achieved, the environment in which it will function should be appropriate for the operators and their tasks. Consideration should be given to noise levels, the thermal environment, lighting and vibration.

Noise levels

'Noise' is any unwanted sound. From a legal standpoint, in order to avoid hearing impairment noise should not exceed 85 dB(A) during a working day. Hearing impairment is not the only reason that noise control is important. Excessive noise can affect concentration and attention and can impede communication between operators. In addition, warning sounds such as truck horns or fire alarms can be masked. Noise can also act as a general stressor.

Noise can be controlled at source in a variety of ways, including: placing heavy, vibrating equipment on a separate, rigid structure; using sound-isolating joints between vibrating equipment and the floor; using vibration isolation mounts; and using damping materials, mufflers and acoustic screens.

If the intensity of sound cannot be controlled at source, then operators should be provided with some form of protection such as ear defenders or ear plugs. The fact that some defenders and plugs are not selective in reducing noise exposure – and may reduce an operator's ability to hear warning sounds – should be taken into account.

Thermal environment

The thermal environment can result in a decrease in the quality of performance if it does not suit the operator and the task they are completing. If it is too hot or cold operators can become irritable, lose concentration and become uncomfortable through perspiring or shivering.

As physical activity tends to generate body heat, consideration should be given to the type of work being completed before temperatures are set in the environment. If operators are involved in heavy manual work they will require a cooler environment than sedentary workers (eg keyboard users). As temperatures start to dip below 16°C, manual dexterity will be impaired. This can eventually contribute to the development of ULDs and operators should therefore have appropriate protective clothing.

Humidity (the level of moisture in the air) and air flow will also influence operator well-being and should be controlled as far as possible. Adjustments to temperature or humidity settings may be required as more people move into an area and as further equipment is introduced.

Computer-controlled temperature settings should be monitored carefully as they tend to work on 'averages' in an area. For example, the average temperature in an office may be 20.5°C but this may be maintained by blasts of cold air through two air vents located in the corners of the room. This may prove to be unacceptable to the operators who sit at these points.

Lighting

Lighting at an appropriate level will assist in maintaining acceptable levels of quality and performance as well as avoiding operator discomfort (eg headaches and eye fatigue). The required level of illuminance – the amount of light falling onto a surface – should be determined by the task demands (eg visual display unit (VDU) work requires a lower lighting level than a written or inspection task). Luminance – the amount of light emitted from a surface – should also be controlled as excessive luminance can cause glare. Glare should be eliminated as far as possible from the environment due to its effect on visual performance.

The location of the light should also be considered. Siting lights on the ceiling or walls may not be the best option. Again, the task requirements will dictate where lights should be located.

Vibration

Vibration is associated with the development of ULDs and back injuries and operators should be protected from it as far as possible. Operators can be exposed to whole and part body vibration through the use of vibrating tools, when driving trucks and from working with and leaning against vibrating machinery. The methods discussed above for reducing noise levels will also have the effect of reducing vibration in many cases.

IN SUMMARY

Mock-ups and trials

- Before full introduction of a new design into the workplace it should be tried out, with as many operators as possible testing the various settings.

Layout

- Items used regularly should be located close to the operator to avoid regular or sustained reaching.

Training

- All operators should have thorough training in the use of their new workstation and/or equipment.

Supervision

- Supervisors should ensure that operators continue to use their new workstation or equipment in the intended manner.

Maintenance and housekeeping

- Designs should not be permitted to deteriorate over time.

Environmental conditions

- Environmental factors such as noise, lighting, heating and vibration will all influence the acceptability of a new design. They should be set at levels which suit the operator and match the requirements of the task.

REFERENCES AND FURTHER READING

Bittner A C and Champney P C (eds), 1995, *Advances in industrial ergonomics and safety VII*, Taylor & Francis, London.

Collingsworth J and Rehahn A (eds), 1993, *Design for disability: a handbook for students and teachers*, London Guildhall University, London.

Corlett E N and Clark T S, 1995, *The ergonomics of workspaces and machines*, Taylor & Francis, London.

Grandjean E, 1988, *Fitting the task to the man* (4th edition), Taylor & Francis, London.

Health and Safety Executive, 1994, *A pain in your workplace? Ergonomic problems and solutions*, HS(G)121, HSE Books, Sudbury.

Health and Safety Executive, 1995, *Sound solutions*, HS(G)138, HSE Books, Sudbury.

Helander M, 1995, *A guide to the ergonomics of manufacturing*, Taylor & Francis, London.

Jones J C, 1992, *Design methods*, Van Nostrand Reinhold, New York.

Pheasant S, 1987, *Ergonomics – standards and guidelines for designers*, Taylor & Francis, London.

Pheasant S, 1996, *Bodyspace – anthropometry, ergonomics and the design of work* (2nd edition), Taylor & Francis, London.

Roebuck J A, 1995, *Anthropometric methods: designing to fit the human body*, Human Factors and Ergonomics Society, Santa Monica CA.

Sheridan T B (ed), 1995, *Analysis, design and evaluation of man-machine systems*, Elsevier Science, London.

CASE STUDY

A sewing machine area within a car assembly plant had a steadily increasing number of ULD cases among the machinists who made up seat covers. As the company in question did not have in-house ergonomists, it decided to enlist outside assistance to deal with the problem.

A thorough task analysis was carried out to evaluate the task demands. During this process, machinists undertaking a range of varied machining tasks were videoed. In addition, they were observed and their activities recorded on a data sheet which provided a breakdown of task components and duration of each sub-component. The machinists were interviewed individually so that they could relay any information on particular activities they completed on specific days and which had not been observed to date. The individual interviews also acted as a means of highlighting any 'personal' styles of work.

At the end of this task analysis, the machinists' work rate, the order in which sections of fabric were stitched together, the number and weight of fabrics being handled, and the postures required to complete the operations were clearly defined. It became apparent from the task analysis and operator interviews that the particular fabrics used to construct the seat covers had certain inherent characteristics which made the sewing operation more difficult.

The workstation design was assessed by comparing its dimensions and layout with the anthropometric data of the user population and with the requirements of the task identified during the task analysis. It was determined that the poor postures adopted by the operators during their machining activities were a direct result of the unsuitable design and dimensions of the workstation.

The work organisational issues, such as rest breaks and rotation, were discussed with managers, supervisors and operators to ensure that there were no serious gaps between the work routine detailed at management level and that which was active at shop floor level. As a result of this exercise it became apparent that certain 'slower' machinists were working through their breaks in an attempt to meet their daily targets. This practice had gone unchecked by the company.

The machinists completed 'comfort' questionnaires which were intended to highlight specific areas – such as wrists, fingers and forearms – which were a source of discomfort. The questionnaire was also designed to assist in the development of an accurate profile of exactly how many operators were suffering from symptoms of a ULD. It could not be assumed that every operator who had developed symptoms had reported this to their supervisors.

At the time when the ULD problem surfaced, there were rumours that the sewing machine area of this plant was to be closed down and some of the operators made redundant. Therefore, operators did not want to draw unnecessary attention to themselves. Using a confidential questionnaire enabled them to be honest about their condition.

Although it was recognised that other factors had contributed to the development of ULDs among the machinists, the workstation design was of greatest concern.

Workstation design

The operators were required to use sewing machines which were attached to the surfaces of workbenches. Each workbench had been constructed of wood and structured so that it had five sides – left and right panels, a back panel, an upper surface and a floor panel which mirrored the upper surface in length, width and depth. In effect, the workstation took the form of a box with the front panel missing. The workstation was of a fixed height, with a Formica-type surface and angled edges. Each sewing machine was located centrally on the worksurface.

Two foot pedals were in use. One was shaped like the sole of a foot and was intended to be used by the right foot to raise and lower the machining foot. The second pedal, which operated the machine, was located to the left of the first and was large enough to be used by both feet. The pedals were attached to the floor panel, at a distance from its leading edge. The right hand pedal was fixed slightly in front of the left and was at a different angle. Neither pedal was adjustable for height, inclination or distance from the operator.

The operators were provided with very basic seating for use at their machines. Many of them were using stools. Some of these stools were identical to those which would typically be found in a bar, while others were the original cast iron 'Singer' stools. None of the chairs or stools were adjustable in any way, most of them did not have any form of seat padding, and the stools obviously did not have any form of back support. In addition, the location of the chairs and stools in relation to the worksurface was restricted by the floor panel which was approximately 80 mm in height. This prevented the chairs and stools being moved forwards, thereby stopping the operators from positioning their knees and legs under the worksurface.

As a result of the workstation design, all operators:

- sat at a distance from the focal point of the sewing machine, ie the needle, because the floor panel prevented them from moving their chairs or stools closer;
- leaned to the left when using the machine because it had been lined up centrally on the worksurface, despite the fact that the needle is located on the left side of the machine;
- had to sit at heights dictated by the particular chair or stool they had been given, which typically forced their arms upwards towards the worksurface;
- did not have any form of back support as a result of either using a stool or having to lean forwards to reach the worksurface;
- had to locate their feet at heights, distances and in orientations dictated by the pedal positions;
- spent long periods leaning on the sharp edge of the worksurface as they fed each piece of material past the needle; and

- struggled to pull the largest pieces of fabric back towards them having fed them past the needle (this was because as each large piece was stitched, it would move over the worksurface and fall over the rear edge – and as the rear edge was sharply angled like the front, this caused the material to become snagged, making its retrieval more difficult).

Workstation redesign

The worksurface was redesigned completely. Its new form was dictated by the needs of the operator and the requirements of the task. The new worksurface was designed to be adjustable in height. The means of adjustment were based on a modified, standard adjustable office desk. The adjustment was achieved by pulling a crank handle out from beneath the worksurface and turning it until the surface approached the desired level. The degree of adjustment available had been determined by evaluating anthropometric data and calculating ranges of dimensions which would be representative of the user population. Once the handle was pushed back into position under the worksurface, the workstation was locked in position and was incapable of being moved unintentionally.

The worksurface was fabricated like a standard office desk and was contoured so that it took the shape of a horseshoe which effectively enveloped the operator. This shape enabled the operator to move closer to the needle and provide support for the arms where necessary. The worksurface was extended in length on the left side to support the weight of large pieces of fabric as they were fed towards the needle. The sewing machine was located so that the needle was positioned centrally on the worksurface, thereby eliminating the need for the operator to lean to the left. The edges of the worksurface were rounded to minimise compression of the wrist. A rollbar was fitted to the rear of the workstation to assist in the retrieval of large pieces of fabric which had fallen over the rear section. It was designed to rotate as the fabric was pulled across it.

The pedals were redesigned so that both were located side by side at the same height, angle and distance from the operator. The pedals were adjustable for height, inclination and distance from the leading edge of the workstation. The degree of movement had been determined by evaluating the appropriate anthropometric data and calculating ranges of dimensions which would be representative of the user population.

Operators were provided with inexpensive standard office chairs (without castors for safety reasons). Each chair was adjustable for height and the backrest was adjustable for height and inclination and incorporated a distinct lumbar support.

Operators were also given new trolleys which had a dual function. Previously, the operators remained at their workstations throughout the course of a shift, except for breaks. Their material and threads were brought to them by another operator who was solely responsible for the replenishment of materials. Therefore, the machinists had no reason to leave their workstations at regular intervals. The trolleys were provided so that operators could collect their own batches of work materials and wheel them back alongside their workstation. The collection of materials provided operators with an opportunity to have a short break from their work and workstation. The trolley's secondary function was to act as an additional worksurface for the storing of materials alongside the workstation.

Trial period

Prior to the full introduction of new workstations for all the machinists, a prototype was built. It was adjustable but on a more basic level than the final design. This prototype was used by as many of the operators as possible under normal working conditions during their shifts. Each operator was given a full briefing on the use of the workstation, and all were encouraged to adjust the pedal and worksurface positions through a wide range of settings. Allowances were made by management for decreases in operator output during these trials.

Operators were videoed during the trials and each of them filled in questionnaires at regular intervals during the use of the prototype. Any verbal, 'throw-away' comments were recorded.

As a result of the trials, which lasted several weeks, ranges of movement were standardised and any 'teething' problems dealt with before the design reached its final form.

Introduction of the new design

Prior to the delivery of all the new workstations, the operators were fully briefed. Once the workstations were installed, every operator was given one-to-one tuition in their use. Production targets were reduced for a period of time to allow the operators to become accustomed to the new way of working. Targets were increased gradually until they reached their original level. This was an essential aspect of making the exercise work – the operators needed to become accustomed to the new postures and efforts required. If such an acclimatisation period had not been provided, operators may have presented with further symptoms of ULDs.

Operators were subsequently monitored on a regular basis following the introduction of the new workstation designs to ensure that they did not encounter problems at a later date.

As a direct result of the alteration in workstation design, the incidence of ULDs was immediately capped and gradually reduced to a minimum. In addition, the level of absenteeism was reduced to an all-time low for that particular work area. This had an obvious positive effect on productivity. Another positive outcome was an increase in the quality of the seat covers being assembled. As a result of the operators being able to sit comfortably when working and being able to reach easily to the appropriate areas of their workstation, they were able to stitch the covers properly at the first attempt. This meant a significant reduction in re-work time.

ANTHROPOMETRICS

Anthropometrics is the branch of ergonomics that deals with the physical size and shape of people. Applying anthropometric data to the design of products and workplaces can go a long way towards ensuring that they will be suited to the physical needs of the end users. This chapter offers an overview of anthropometrics and is designed to convey the importance of this key area. As an introduction to the value of anthropometric data, some commonly used dimensions are provided for British adults aged from 19 to 65 years.

While there are various anthropometric data sources available, it is often difficult to find one that is fully suitable. Anthropometric surveys often relate to an irrelevant user group, do not contain the dimension specifically required, or were undertaken on a commercial 'in confidence' basis and are therefore not in the public domain. Alternatively, surveys may be too old to be reliable. Undertaking large scale anthropometric surveys is a time-consuming and expensive activity. Consequently, data may not be available for a number of groups in which the ergonomist may have an interest. It is an inescapable fact that more data are derived from military sources (particularly the United States military) than for any other occupational group. While these data may be invaluable to those ergonomists operating in the military arena, they are of limited use in the design of a factory to be used solely by a UK civilian workforce. For example, the military group is likely to be younger, stronger, lighter, taller and have fewer females than a cross-section of factory workers. Before applying data to a particular situation it is important that the ergonomist is confident that they are up-to-date and relevant.

It is generally thought that each subsequent generation is larger than its predecessors. However, there is evidence to suggest that some societies are 'bottoming out'; this is particularly true of the more affluent societies. Broadly speaking, a person's size and shape are primarily dependent on inherited genetic material, diet and lifestyle. Higher standards of living in many parts of the world have resulted in people enjoying substantially improved diets and this may go a long way towards explaining why some of the older databases may not be appropriate for designing for the people of today – or tomorrow.

It is often the case that designers, engineers and others will create products and environments that appear to be suited only to themselves – or to the 'average' person. This is misguided. For example, if the height of a doorway were set to suit the average stature, 50 per cent of the population would need to duck in order to gain access to the room. It is obvious that the door needs to be considerably higher than average – the question is 'how high'? At the time of writing, the UK's tallest man is Chris Greener at 2,292 mm – 7' 6¼". But is it realistic to design every door to suit him?

The largest dimensions are not, in any case, always of significance. For example, the design of a machinery guard for a piece of equipment to which only adults will have access would allow child data to be discarded, allowing the focus to be on the smallest adult finger, hand and arm on the basis that if those with the smallest limbs cannot move their hand past the guard then others will be protected automatically.

Therefore, designers and others should aim to accommodate a range of sizes; how great the range should be will depend largely on how critical it will be to exclude parts of the population. Clearly, in situations where an exclusion may lead to a catastrophic outcome, for example, where a pilot cannot reach the controls crucial to keeping a plane airborne, then the burden on the designer to cater for a wider range is greater than in other situations, such as trying to establish the acceptable height for a park bench.

What, then, would be a reasonable range for a designer to work with for most non-critical situations? If a researcher selected the centre of a large town and proceeded to measure the stature of a sufficient number of passers-by, it is likely that data would be produced such as those used to plot the distribution curve shown below. This curve is known as a normal or Gaussian distribution and it illustrates that the likelihood of encountering people who are of average or near to average height is far greater than encountering someone who is exceptionally tall or small.

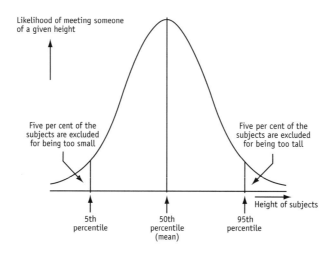

The normal distribution

It has been traditional for designers to exclude the physical extremes on the basis of rarity. It is probably more honest to say that they are excluded because trying to design for everyone would typically be too great a challenge; sometimes without a solution. The point at which the exclusion begins is somewhat arbitrary, however, it is widely accepted that the cut off points are set at what are known as the 5th and 95th percentiles. This means that five per cent of the population is, for example, too tall to be catered for by the design, and five per cent of the population is, for example, too short and is similarly excluded. Hence, 90 per cent of the population is included in the design philosophy and, hopefully, their needs will be met. When focusing on a population of potential users, the hallmark of a good design will be the inclusion of as many people as practicable over the 90 per cent threshold.

Applying anthropometric data

The data used in the remainder of this chapter are adapted from 'Bodyspace – anthropometry, ergonomics and the design of work' (Pheasant 1996). The data, anthropometric estimates for British adults aged from 19 to 65 years, have been selected as those likely to be of most general use to the reader.

Before applying data from any source it should be established that the most appropriate data have been accessed. It should also be noted that most anthropometric data are collected from subjects who are either nude or near nude. This means that 'corrections' need to be made for clothing (which will probably vary between seasons), shoes and any other items such as breathing apparatus. The postures adopted by people during the measurement process tend to be somewhat unnatural and more erect than would be the case generally. Therefore, how much people 'slump' – relax into a more natural and sustainable posture – may have to be considered. The tables that appear later in the chapter offer some estimates for how much 'correction' may be needed to adjust the dimensions given to allow for 'slumping'. As long as those using the data are aware of how they were collected the appropriate modifications can be made with comparative ease.

Example

Using anthropometric data to define the height of a worksurface on which a light, manipulative task will be undertaken

If operators can adopt an upright, forward-facing posture while working they will be less likely to develop back problems. The further the individual deviates from an upright posture, the more stressful it is for the lower back. Equally, working above elbow height can cause upper limb and shoulder problems. The working height created by the combination of the workstation and the tasks undertaken largely determines the upper body posture adopted by the operator. To allow operators to remain in an upright position they should be presented with a working height which is at or slightly below their elbow height when standing (or when seated). This is measured with the elbows held close to the body and arms bent at 90°. The upper arm should hang from the shoulder in a relaxed position and the forearm should be horizontal or declined so that the hands are slightly below elbow height.

Most people find it more comfortable to sit rather than stand. Irrespective of the arguments for standing (eg improved lumbar lordosis, lower limb circulation and muscle tone), the fact is that sitting is the preferred option for most operators as it is generally less tiring. Increasingly, employers are taking the view that there are merits and de-merits to both options, and that they should provide operators with opportunities for sitting and standing. There is much to commend this approach and ideally both options should be available at all appropriate workstations.

The standing workstation

Since no two individuals will have exactly the same body proportions and a number of different operators may use a particular workstation, it is preferable to offer some aspect of adjustability to cater for these variations. However, adjustability may not be viable; the following discussion assumes that a fixed height surface will be employed.

How high should a worksurface be so that an operator can stand on the floor and work at it safely? When dealing with the working height, it is usual to consider the height of the worksurface plus any additional height for the product – or the height at which the operator's hands will be as they work. This assumes that there are no guards or other barriers that the operators have to negotiate to gain access to the products they manipulate. For the purposes of this exercise, it should be assumed that the product has no appreciable thickness.

The calculations will try to accommodate, as a minimum, all users from small females (5th percentile) to large males (95th percentile). As discussed earlier, it is better for an operator to work at or slightly below elbow height. The standing elbow height of a 5th percentile female is 930 mm plus an allowance of 25 mm for her shoes, giving a total of 955 mm (the shoe allowance for females used here is conservative but based on the knowledge that higher-heeled footwear is unusual in an industrial setting). Therefore, the worksurface height for a small female undertaking a lightweight manipulative task on a product with no appreciable thickness would be 955 mm.

The 95th percentile male has a standing elbow height of 1,175 mm plus 25 mm for his shoes, giving a total of 1,200 mm for his worksurface height. This is 245 mm above the height established as preferable to fit a small woman. The standing knuckle height (the height of the knuckles above floor level when standing with the arms hanging down relaxed at the side of the body) of a 95th percentile male is 822 mm plus 25 mm for his shoes, giving a total of 847 mm. Undertaking any hand-based task below this height would oblige him to bend over/down. Therefore, although a large man will be working well below his preferred working height (ie his elbow height) when using the 955 mm high worksurface already established to 'fit' the 5th percentile female, he would still be working above his knuckle height. His working position would not be optimal but it would be satisfactory and he would not need to bend over/down to reach the worksurface, providing that the work he was undertaking did not require him to reach forward.

If it was decided to raise the worksurface height simply to satisfy the preferences of the larger men, problems would be encountered in accommodating the smaller females because they would be obliged to elevate their shoulders and upper limbs to reach the surface and this is known to be fatiguing and hazardous. These difficulties could be overcome by the introduction of platforms of various sizes so that the female operators could elevate themselves to maintain good postures. However, storage and distribution issues – as well as fall and trip hazards potentially introduced by such platforms – could make them impractical.

No single height is equally acceptable to everyone – 955 mm is probably as close as possible to a 'universal' height although, in reality, it suits the smaller females better than the larger males. Any notion of 'averaging' the dimensions should not be entertained; the advantages to the taller men would be greatly outweighed by the implications for the smaller women. Obviously, no operator would be placed at a disadvantage if an adjustable height worksurface were available. However, it is more usual to provide adjustable seating as a means of overcoming variances and improving operator fit.

The seated workstation

The principles outlined above with regard to working elbow height are effectively the same for the seated operator. Thus, a correctly seated 5th percentile female will be supported by a seat surface that is 381 mm above the floor. This is calculated by adding her popliteal height (the height from the floor to the back of her knee which approximates to seat height in bare feet) of 356 mm to her heel size of 25 mm. If her elbow to seat dimension of 187 mm is added, a seated elbow height of 568 mm is established, which represents the 'ideal' worksurface height for this operator.

Applying the same principles, the 95th percentile male will be sitting 513 mm above the floor with elbows at 809 mm, which represents his 'ideal' worksurface height.

These values give the worksurface heights that the two 'extremes' can safely work to, providing that they are seated with their feet flat on the floor.

Assuming that the worksurface to be sat at in this example has been set at the 955 mm established earlier to allow operators to stand, the appropriate seat heights for all likely operators using this workstation are calculated as follows:

	5th percentile female	95th percentile male
original seat height (feet on floor)	381	513
original seated elbow height (feet on floor)	568	809
the worksurface height to be sat at is	955	955
therefore, to reach the worksurface, the seat must be raised by	387 (955 - 568)	146 (955 - 809)
'new' seat height	768 (381 + 387)	659 (513 + 146)
a footrest height equivalent to the height that the seat has been raised to must be provided to support the feet	387	146

All dimensions are in millimetres

Thus, a seat that adjusts 659–768 mm with a footrest or footring that adjusts 146–387 mm will allow these operators to adopt an appropriate seated posture at a 955 mm high worksurface.

Clearly, if the worksurface were lower or higher, the height difference value could be subtracted or added to these figures. For example, if the worksurface height were 100 mm higher than in the example used here then the seat and footrest would need to be 100 mm higher across the adjustment range. However, it should be borne in mind that while this would satisfy the need for a desirable seated posture it would not allow most operators to stand. The 955 mm high worksurface does allow operators, albeit with varying degrees of movement away from the 'ideal', to stand as well as sit at the worksurface.

Calculating the percentage of people not accommodated by a worksurface height

The normal distribution curve (discussed earlier) for a population of users can be described by reference to two parameters: the mean or average and the standard deviation. The mean is the centre of the distribution and the standard deviation is an index of the degree of variation about the centre. Many reference texts on anthropometrics give the population mean and the standard deviation, along with a few selected percentiles – typically the 5th, 50th and 95th. These are all the data needed if the design is shown to accommodate the whole range. However, sometimes constraints are placed on the design and it does not satisfy the full range. It can then be helpful to know exactly what percentage of the population has been excluded. Where the mean and the standard deviation for a given distribution are known, any percentile can be calculated with the equation:

$$p = m + zs$$

where '**p**' is the dimension for a specific percentile

'**m**' is the known mean

'**z**' is the 'z' score – a statistical constant for a given percentile*

'**s**' is the known standard deviation

*'z' scores are reproduced in the table opposite. The statistical basis for the use of 'z' scores is beyond the scope of this book. Since 'z' scores are constants, they can be applied to any data where the mean and standard deviation are known (see pages 47–56).

Use of the equation allows the generation of a complete set of data for any relevant body dimension. An example of a dataset for standing elbow height is shown in the table opposite. This dataset was created by inserting the following data (given on page 48), along with the appropriate 'z' scores, into the formula:

standing elbow height	male	female
50th percentile (mean)	1090	1005
standard deviation	52	46

All dimensions are in millimetres

The easiest and quickest method of generating a data table is to use a computer spreadsheet and copy the formula from one cell to another.

Tables that show the intermediate percentiles are very useful. For example, if engineering constraints dictate that the lowest practicable height for a worksurface is 1,050 mm, it would be helpful to establish what percentage of the user population would be prejudiced. For a person to work ideally at the 1,050 mm worksurface they would need a standing elbow height of 1,050 mm. The shoe heel height of 25 mm should be subtracted to establish the actual elbow height. Therefore, the required elbow height is 1,025 mm. A person of either sex who has a standing elbow height of 1,025 mm would be able to stand at the 1,050 mm worksurface in 25 mm high shoes. The data in the table opposite indicate that 10 per cent of males and 67 per cent of females have a standing elbow height under 1,025 mm and would therefore not be able to stand and work at a 1,050 mm high worksurface without elevating their arms. Excluding such a large percentage of individuals would make this height unacceptable.

If such a table is not available, there is an alternative method for arriving at the same answer when a specific dimension is under consideration. The formula **p=m+zs** can be manipulated to give:

$$\frac{p-m}{s} = z$$

Thus, for the 1,050 mm worksurface example, the actual elbow height of 1,025 mm (percentile value 'p' in the formula) would be inserted into the alternative formula, as follows:

male $\dfrac{1025 - 1090}{52} = z$ female $\dfrac{1025 - 1005}{46} = z$

$-1.25 = z$ $0.43 = z$

Data for standing elbow height (British adults aged 19–65 years)

Standard deviations: 52 (males) and 46 (females)

%ile	z score	males	females	%ile	z score	males	females
1	-2.33	969	898	51	0.03	1092	1006
2	-2.05	983	911	52	0.05	1093	1007
3	-1.88	992	919	53	0.08	1094	1009
4	-1.75	999	925	54	0.10	1095	1010
5	-1.64	1005	930	55	0.13	1097	1011
6	-1.55	1009	934	56	0.15	1098	1012
7	-1.48	1013	937	57	0.18	1099	1013
8	-1.41	1017	940	58	0.20	1100	1014
9	-1.34	1020	943	59	0.23	1102	1016
10	-1.28	1023	946	60	0.25	1103	1017
11	-1.23	1026	948	61	0.28	1105	1018
12	-1.18	1029	951	62	0.31	1106	1019
13	-1.13	1031	953	63	0.33	1107	1020
14	-1.08	1034	955	64	0.36	1109	1022
15	-1.04	1036	957	65	0.39	1110	1023
16	-0.99	1039	959	66	0.41	1111	1024
17	-0.95	1041	961	67	0.44	1113	1025
18	-0.92	1042	963	68	0.47	1114	1027
19	-0.88	1044	965	69	0.50	1116	1028
20	-0.84	1046	966	70	0.52	1117	1029
21	-0.81	1048	968	71	0.55	1119	1030
22	-0.77	1050	970	72	0.58	1120	1032
23	-0.74	1052	971	73	0.61	1122	1033
24	-0.71	1053	972	74	0.64	1123	1034
25	-0.67	1055	974	75	0.67	1125	1036
26	-0.64	1057	976	76	0.71	1127	1038
27	-0.61	1058	977	77	0.74	1128	1039
28	-0.58	1060	978	78	0.77	1130	1040
29	-0.55	1061	980	79	0.81	1132	1042
30	-0.52	1063	981	80	0.84	1134	1044
31	-0.50	1064	982	81	0.88	1136	1045
32	-0.47	1066	983	82	0.92	1138	1047
33	-0.44	1067	985	83	0.95	1139	1049
34	-0.41	1069	986	84	0.99	1141	1051

All male and female data are in millimetres

continued...

%ile	z score	males	females	%ile	z score	males	females
35	-0.39	1070	987	85	1.04	1144	1053
36	-0.36	1071	988	86	1.08	1146	1055
37	-0.33	1073	990	87	1.13	1149	1057
38	-0.31	1074	991	88	1.18	1151	1059
39	-0.28	1075	992	89	1.23	1154	1062
40	-0.25	1077	994	90	1.28	1157	1064
41	-0.23	1078	994	91	1.34	1160	1067
42	-0.20	1080	996	92	1.41	1163	1070
43	-0.18	1081	997	93	1.48	1167	1073
44	-0.15	1082	998	94	1.55	1171	1076
45	-0.13	1083	999	95	1.64	1175	1080
46	-0.10	1085	1000	96	1.75	1181	1086
47	-0.08	1086	1001	97	1.88	1188	1091
48	-0.05	1087	1003	98	2.05	1197	1099
49	-0.03	1088	1004	99	2.33	1211	1112
50	0.00	1090	1005				

When the calculated 'z' scores are located in the table, it can be seen that the same percentages of males and females are excluded as calculated with the original formula.

This is a useful formula for quickly deciding whether a particular dimension is acceptable or not. Its drawback is that, unlike a full dataset, it does not show the alteration required to the design in order to reach an acceptable solution. For example, a full data table shows, at a glance, that lowering the worksurface height by another 30 mm to 1,005 mm would make it acceptable to 95 per cent of men but still exclude 50 per cent of women. This may make the worksurface height acceptable if it will be used exclusively by men. Relative changes such as this can be checked out quite simply by using a full table of data rather than making a series of single calculations.

Anthropometric estimates for British adults aged 19–65 years

The data that follow are presented with an indication of how they were measured and an example of what they might be used for. Because the mean is given along with the standard deviation, it is possible to calculate all the intermediate percentiles, as described earlier. The 5th and 95th percentiles are given as convenient benchmarks and 'correction' factors are listed to improve the accuracy of the dimensions. The 'z' scores given in the above table are statistical constants and can be applied to all the data.

This dataset is offered as a useful tool for making assessments in a workplace setting because it covers British adults of working age. It will not be appropriate for all environments – for example, if a workforce consists mainly of teenagers. Alternative datasets should be used in such cases.

ANTHROPOMETRICS

STATURE	**MALE**				**FEMALE**			
	5th %ile	50th %ile	95th %ile	SD	5th %ile	50th %ile	95th %ile	SD
	1625	1740	1855	70	1508	1610	1712	62

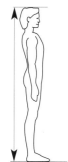

Measured between: the crown of the head and the floor for a person standing 'bolt upright' and without shoes

Used for: clearance required for overhead obstacles

Corrections: shoes (males 25 mm, females 45 mm), hats (25 mm), helmets (35 mm)

STANDING EYE HEIGHT	**MALE**				**FEMALE**			
	5th %ile	50th %ile	95th %ile	SD	5th %ile	50th %ile	95th %ile	SD
	1517	1630	1743	69	1405	1505	1605	61

Measured between: the corner of the eye to the floor

Used for: the reference level for the design and location of visual displays when standing

Corrections: shoes (males 25 mm, females 45 mm)

STANDING SHOULDER HEIGHT	**MALE**				**FEMALE**			
	5th %ile	50th %ile	95th %ile	SD	5th %ile	50th %ile	95th %ile	SD
	1317	1425	1533	66	1215	1310	1405	58

Measured between: the top of the bony projection near to the edge of the shoulder (the acromium) and the floor

Used for: approximately defining the centre of rotation of the shoulder joint. Used to establish comfortable reach envelopes. Any frequently used control should not exceed this height for a standing operator

Corrections: shoes (males 25 mm, females 45 mm)

All dimensions are in millimetres

ANTHROPOMETRICS

STANDING ELBOW HEIGHT	MALE				FEMALE			
	5th %ile	50th %ile	95th %ile	SD	5th %ile	50th %ile	95th %ile	SD
	1005	1090	1175	52	930	1005	1080	46

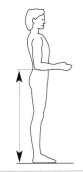

Measured between: the bony projection on the underside of the elbow and the floor

Used for: the key reference level for determining the height of a worksurface for a standing operator

Corrections: shoes (males 25 mm, females 45 mm)

KNUCKLE HEIGHT	MALE				FEMALE			
	5th %ile	50th %ile	95th %ile	SD	5th %ile	50th %ile	95th %ile	SD
	688	755	822	41	661	720	779	36

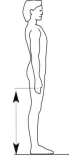

Measured between: the knuckles and the floor

Used for: establishing the lowest acceptable level for most types of hand operated control

Corrections: shoes (males 25 mm, females 45 mm)

SITTING HEIGHT	MALE				FEMALE			
	5th %ile	50th %ile	95th %ile	SD	5th %ile	50th %ile	95th %ile	SD
	851	910	969	36	793	850	907	35

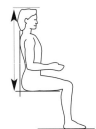

Measured between: the crown of the head and the seat plane for a person sitting 'bolt upright'

Used for: establishing the clearance required for overhead obstacles for a seated person

Corrections: hats (25 mm), helmets (35 mm), heavy clothing (10 mm), seat cushion compression (varies), seating slump (up to 40 mm)

ANTHROPOMETRICS

SITTING EYE HEIGHT	MALE				FEMALE			
	5th %ile	50th %ile	95th %ile	SD	5th %ile	50th %ile	95th %ile	SD
	733	790	847	35	686	740	794	33

Measured between: the corner of the eye and the seat plane

Used for: the reference level for the design and location of visual displays when seated

Corrections: heavy clothing (10 mm), seat cushion compression (varies), seating slump (up to 40 mm)

SITTING SHOULDER HEIGHT	MALE				FEMALE			
	5th %ile	50th %ile	95th %ile	SD	5th %ile	50th %ile	95th %ile	SD
	543	595	647	32	504	555	606	31

Measured between: the top of the bony projection near to the edge of the shoulder (the acromium) and the seat plane

Used for: approximately defining the centre of rotation of the shoulder joint. Can be used to establish comfortable reach envelopes. Any frequently used control should not exceed this height

Corrections: heavy clothing (10 mm), seat cushion compression (varies), seating slump (up to 40 mm)

SITTING ELBOW HEIGHT	MALE				FEMALE			
	5th %ile	50th %ile	95th %ile	SD	5th %ile	50th %ile	95th %ile	SD
	194	245	296	31	187	235	283	29

Measured between: the bony projection on the underside of the elbow and the seat plane

Used for: establishing the height of the armrests on seats

Corrections: heavy clothing (10 mm), seat cushion compression (varies), minimal seating slump

THIGH THICKNESS	MALE				FEMALE			
	5th %ile	50th %ile	95th %ile	SD	5th %ile	50th %ile	95th %ile	SD
	135	160	185	15	127	155	183	17

Measured between: the highest part of the thigh and the seat plane

Used for: establishing how much space is required between a seat and the underside of a worksurface or other obstacle

Corrections: seat cushion compression (varies), heavy outdoor clothing (35 mm)

POPLITEAL HEIGHT	MALE				FEMALE			
	5th %ile	50th %ile	95th %ile	SD	5th %ile	50th %ile	95th %ile	SD
	392	440	488	29	356	400	444	27

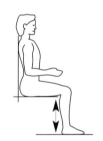

Measured between: the back of the knee and the floor

Used for: defining the maximum height of a seat

Corrections: shoes (males 25 mm, females 45 mm)

BUTTOCK – KNEE LENGTH	MALE				FEMALE			
	5th %ile	50th %ile	95th %ile	SD	5th %ile	50th %ile	95th %ile	SD
	544	595	646	31	521	570	619	30

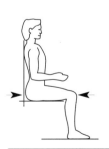

Measured between: the back of the uncompressed buttocks to the front of the knee in the standard sitting position

Used for: minimum distance between seats for access and egress

Corrections: heavy outdoor clothing (20 mm)

ANTHROPOMETRICS

BUTTOCK – POPLITEAL LENGTH	MALE				FEMALE			
	5th %ile	50th %ile	95th %ile	SD	5th %ile	50th %ile	95th %ile	SD
	443	495	547	32	431	480	529	30

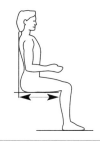

Measured between: horizontal distance from the uncompressed buttock to the popliteal angle at the back of the knee

Used for: determining the maximum acceptable depth (ie front to back dimension) of a seat

Corrections: heavy outdoor clothing (20 mm)

SHOULDER BREADTH (BIDELTOID)	MALE				FEMALE			
	5th %ile	50th %ile	95th %ile	SD	5th %ile	50th %ile	95th %ile	SD
	419	465	511	28	356	395	434	24

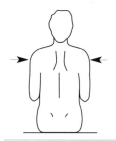

Measurement: maximum breadth across the shoulders

Used for: clearance required in upper part of work space at shoulder level

Corrections: indoor clothing (10 mm), heavy outdoor clothing (40 mm)

SHOULDER BREADTH (BIACROMIAL)	MALE				FEMALE			
	5th %ile	50th %ile	95th %ile	SD	5th %ile	50th %ile	95th %ile	SD
	367	400	433	20	325	355	385	18

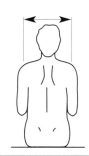

Measured between: the acromi (the bony prominences which form the points of the shoulders)

Used for: establishing the lateral separation of the centres of rotation of the upper limbs and applied when calculating the zone of convenient reach

Corrections: none

ANTHROPOMETRICS

HIP BREADTH	MALE				FEMALE			
	5th %ile	50th %ile	95th %ile	SD	5th %ile	50th %ile	95th %ile	SD
	312	360	408	29	308	370	432	38

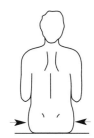

Measured between: the hips at their maximum width

Used for: clearance required at seat height (eg between the arms of chairs). The breadth of the supporting surface of a seat cushion should not be much less than this

Corrections: light clothing (10 mm), medium clothing (25 mm), heavy outdoor clothing (50 mm)

ABDOMINAL DEPTH	MALE				FEMALE			
	5th %ile	50th %ile	95th %ile	SD	5th %ile	50th %ile	95th %ile	SD
	218	270	322	32	206	255	304	30

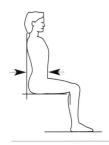

Measured between: the front of the abdomen and the vertical reference plane in the standard sitting position

Used for: the clearance between the seat back and any obstacles

Corrections: light clothing (10 mm), medium clothing (25 mm), heavy outdoor clothing (50 mm)

ELBOW – FINGERTIP LENGTH	MALE				FEMALE			
	5th %ile	50th %ile	95th %ile	SD	5th %ile	50th %ile	95th %ile	SD
	441	475	509	21	399	430	461	19

Measured between: the back of the elbow and the tip of the middle finger in the standard sitting position

Used for: forearm reach and defining the normal working area

Corrections: none

ANTHROPOMETRICS

SHOULDER – GRIP LENGTH	MALE				FEMALE			
	5th %ile	50th %ile	95th %ile	SD	5th %ile	50th %ile	95th %ile	SD
	613	665	717	32	552	600	648	29

Measured between: the acromium and the centre of a bar gripped in the hand when the elbow and wrist are straight

Used for: establishing the functional length of the upper limb and defining the zone of convenient reach

Corrections: none

FORWARD GRIP REACH	MALE				FEMALE			
	5th %ile	50th %ile	95th %ile	SD	5th %ile	50th %ile	95th %ile	SD
	724	780	836	34	654	705	756	31

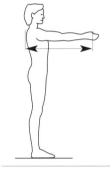

Measured between: the back of the shoulder blades and the centre of a bar gripped in the hand when the elbow and wrist are straight. This is an 'easy' reach undertaken without stretching

Used for: establishing the distance at which items can be gripped, manipulated or operated without undue stretching

Corrections: none

VERTICAL GRIP REACH – STANDING	MALE				FEMALE			
	5th %ile	50th %ile	95th %ile	SD	5th %ile	50th %ile	95th %ile	SD
	1929	2060	2191	80	1789	1905	2021	71

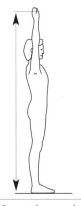

Measured between: the floor and the centre of a bar gripped in the hand when the elbow and wrist are straight while standing. This is an 'easy' reach undertaken without stretching

Used for: establishing the height at which items can be gripped, manipulated or operated without undue stretching

Corrections: shoes (males 25 mm, females 45 mm)

VERTICAL GRIP REACH – SITTING

	MALE				FEMALE			
	5th %ile	50th %ile	95th %ile	SD	5th %ile	50th %ile	95th %ile	SD
	1147	1245	1343	60	1063	1150	1237	53

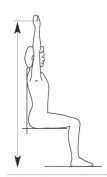

Measured between: the floor and the centre of a bar gripped in the hand when the elbow and wrist are straight while sitting. This is an 'easy' reach undertaken without stretching

Used for: establishing the height at which items can be gripped, manipulated or operated without undue stretching from the seated position

Corrections: heavy clothing (10 mm), seat cushion compression (varies)

ELBOW SPAN

	MALE				FEMALE			
	5th %ile	50th %ile	95th %ile	SD	5th %ile	50th %ile	95th %ile	SD
	868	945	1022	47	779	850	921	43

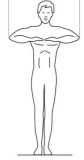

Measured between: the tips of the elbows when both upper limbs are stretched out sideways and the elbows are fully flexed so that the fingertips touch the chest

Used for: a guideline when considering 'elbow room' in the workplace

Corrections: medium outdoor clothing (20 mm)

ARM SPAN

	MALE				FEMALE			
	5th %ile	50th %ile	95th %ile	SD	5th %ile	50th %ile	95th %ile	SD
	1654	1790	1926	83	1489	1605	1721	71

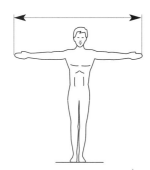

Measured between: the fingertips when both arms are fully stretched out sideways

Used for: establishing maximum lateral reach

Corrections: none

ANTHROPOMETRICS

HEAD BREADTH	MALE				FEMALE			
	5th %ile	50th %ile	95th %ile	SD	5th %ile	50th %ile	95th %ile	SD
	145	155	165	6	135	145	155	6

Measurement: the maximum breadth of the head above the ears

Used for: clearances to avoid trapping

Corrections: add 35 mm for the ears, up to 90 mm for a safety helmet and 120 mm for ear defenders

HAND BREADTH	MALE				FEMALE			
	5th %ile	50th %ile	95th %ile	SD	5th %ile	50th %ile	95th %ile	SD
	77	85	93	5	68	75	82	4

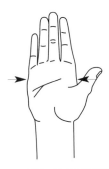

Measurement: the maximum breadth across the palm of the hand

Used for: establishing the size of handles and grips

Corrections: gloves (up to 25 mm)

FOOT LENGTH	MALE				FEMALE			
	5th %ile	50th %ile	95th %ile	SD	5th %ile	50th %ile	95th %ile	SD
	242	265	288	14	215	235	255	12

Measured between: the back of the heel and the tip of the longest toe

Used for: establishing foot clearances under equipment and benches and for pedal layouts and foot operated switches

Corrections: footwear (up to 40 mm depending on the design)

ANTHROPOMETRICS

FOOT BREADTH	MALE				FEMALE			
	5th %ile	50th %ile	95th %ile	SD	5th %ile	50th %ile	95th %ile	SD
	85	95	105	6	80	90	100	6

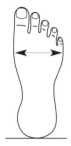

Measurement: maximum horizontal breadth across the foot

Used for: establishing foot clearances for pedal layouts and foot operated switches

Corrections: footwear (up to 30 mm depending on the design)

The use of anthropometric data can be a valuable part of the design process – providing the information is from an appropriate source and applicable to a particular project. However, data can only go so far, and it is recommended that new designs undergo a full mock-up stage involving a structured fitting trial with a representative sample of the target user population.

REFERENCES AND FURTHER READING

Corlett E N and Clark T S, 1995, *The ergonomics of workspaces and machines*, Taylor & Francis, London.

Di Martino V and Corlett N (eds), 1998, *Work organisation and ergonomics*, ILO, Geneva.

Haslegrave C M, 1979, An anthropometric survey of British drivers, *Ergonomics*, **22**, 145–54.

Kroemer K and Grandjean E, 1997, *Fitting the task to the human* (5th edition), Taylor & Francis, London.

Panero J and Zelnic M, 1979, *Human dimension and interior space*, Architectural Press, Oxford.

Pheasant S, 1996, *Bodyspace – anthropometry, ergonomics and the design of work* (2nd edition), Taylor & Francis, London.

DISPLAY SCREEN EQUIPMENT WORK

There is a common misconception that everyday use of computers will result in a change in the quality of an individual's eyesight, or will lead to headaches, backache or possibly an upper limb disorder. However, such problems relate directly to *how* computers are used. They are often put on any available desk, irrespective of its height or surface area. The desks are usually arranged in a way which achieves the maximum number of people and pieces of equipment in the available space without any consideration of the effect the layout will have. The computer user will probably be allocated any available chair and they will probably set to work before adjusting it. In addition, it is common for users to work for long periods without a break.

Many problems could be avoided if computer users were provided with appropriate workstation equipment, and if they used this equipment properly. Of course, this does not just apply to office-based organisations. It applies equally to industrial sites which have their own office complexes, or control rooms which monitor production operations.

Workstation equipment

Before selecting any workstation equipment, consideration should be given to the variability of the individuals who will work in the area and the tasks they will carry out. It is only by recognising the needs of both the individual and the task that an appropriate workstation can be provided. The Health and Safety (Display Screen Equipment) Regulations 1992 and associated guidance and supporting standards – such as BS EN ISO 9241-5 (British Standards Institution (BSI) 1999) – should be referred to.

Seating

The chair is the key to good posture and only suitably designed chairs should be provided. A chair does not need to be sophisticated or expensive to be effective. A simple, well-made chair will be far more useful to office personnel than an extremely complex 'designer' chair which they cannot use properly. Although suppliers may convince purchasers that a particular chair is the best design, they can only be sure of this by matching its design and dimensions against appropriate standards, comparing its characteristics with the user population and the task demands, and using it in the real working situation.

Adjustments
The chair should offer a degree of flexibility through adjustment. Adjustments should be capable of being made easily from the seated position. If the adjustments are difficult to reach and use, users may not adjust the chair and are likely to adopt postures which are unsuitable and which will increase discomfort.

Backrest
The backrest is probably the most important part of the chair. It should be adjustable for forwards and

backwards tilt and, ideally, should have independent height adjustment (ie it should be capable of being moved up and down independently of the seat). Although the Schedule to the Display Screen Equipment Regulations does not require this independent height adjustment, it is considered to be an essential requirement by most ergonomists. This is because of the use of the lumbar support (the soft curve or bump towards the bottom of most backrests). The aim should be to move this support up or down so that it fits snugly into the curve of the lower back. If a chair has been designed with a distinct lumbar support but with no up and down movement, there is no guarantee that the support will actually be in the right position in relation to a user's back. If the user's body dimensions do not 'fit' the chair's design, their lower back will not have adequate support. This problem can only be overcome if the backrest is height-adjustable, or if the lumbar support is capable of being moved up and down inside it, such as in some car seats.

Armrest

Many users feel 'hard done by' if they are given a chair without armrests. However, they are probably at an advantage as a result of not having to cope with them. The majority of chairs in offices are provided with fixed armrests. They are either bolted or welded onto the underside of the chair and cannot be moved in any direction. The main problem is that they usually act as an obstruction, preventing the user from being as close to the worksurface as they would want. As a consequence, the user has to either lean forwards to reach the keyboard, or perch on the front edge of the chair. Both these sitting positions result in a total loss of back support. Some users lower the seat to move the armrests under the desk which causes them to sit at too low a level for their screen and keyboard work.

The height and width settings of fixed armrests will not suit all potential users. Some people will find them too low and will lean to one side, or 'slouch' in the seat to rest their arm(s) on them. If they are too high, the user's shoulders will be elevated as they lift their forearms onto their surfaces. If they are too far apart, users will move their elbows away from their bodyside to reach them, and if they are too close together, larger users will find them uncomfortable. Ideally, armrests should be adjustable for both height and width settings so that users can move them to a position where they support the forearms comfortably. Users should not have to 'work' to use an armrest.

If fixed armrests are proving to be a problem for users, they should be removed where practicable.

Five star base

A five star base (the five feet on the base of the chair) is a requirement of the Schedule to the Display Screen Equipment Regulations. It gives the user stability, preventing them from tipping up when they lean or reach forwards. Castors are the norm on the base of task chairs. However, if they are used on a vinyl or concrete floor, such as in an industrial or workshop setting, they may prove to be hazardous (the slightest movement by the user may result in the chair racing freely across the smooth surface). Glides may be a more appropriate alternative.

Certain chairs are available with six or more castors. These offer increased stability and are particularly useful if the user is heavily overweight or is a wheelchair user. The overweight individual will rely on the arms of a chair to manoeuvre themselves in and out of the seat, and the wheelchair user will use the armrests for total support as they manoeuvre themselves out of their wheelchair and into the task chair. It is at such times that a chair will be susceptible to tipping – and a six star (or more) base will reduce the likelihood of this occurring.

Backless chairs

Some people use backless chairs at their desks and, if they have a health condition which necessitates their use, these chairs can contribute to their comfort levels. However, the design of backless chairs can cause additional problems for users. The seat is tilted forwards with the aim of ensuring that the user adopts an upright S-shaped spine. However, as the chair is backless the user's back is not supported while they are working. In addition, the user leans on knee pads which are an integral part of the chair's design. The knees are not designed to support the bodyweight for long periods and as a result they develop pressure points and become uncomfortable.

Desks

A desk should not be viewed simply as a surface on which to rest objects. There are many aspects to consider before deciding that a desk is suitable for use and introducing it to the working environment.

Height

The height of the worksurface will determine, in part, the posture adopted by the user. The 'standard' height for a non-adjustable desk is 720 mm, measured from floor level to the upper surface. The supplier will probably claim that this is acceptable for all users. However, this is unlikely to be the case for the tallest of users who will probably spend much of their day leaning forwards over their desk, experiencing some degree of backache. The taller user may have to be provided with a higher desk, so that they can adopt a more upright posture (in the short term, the desk could be raised, securely, on blocks). It is important to build a profile of the user population and select desks to suit specific needs.

At the other end of the scale are smaller than average users who will almost always find that 'standard' desks are too high for them to use comfortably. This problem can be overcome by adjusting the chair to raise the user and, if necessary, providing a footrest.

Providing users with adjustable height desking should, in theory, eliminate possible problems with size, as heights can then be adjusted to suit individual dimensions. The cost would not necessarily be prohibitive – there is often little difference between fixed and adjustable height desking, within individual product ranges.

Surface area

One of the biggest problems faced by display screen equipment (DSE) users is that the surface area of the desk is not big enough for all their equipment and documents. As a consequence, users are often forced to site their screen, keyboard, telephone and so on where they will fit on the desk surface, as opposed to where they should be located. It is quite common to see a screen in the left or right corner of the desk with the user having to look sideways when referring to it, often resulting in neckache. The head weighs around 5 kg, hence neck and shoulder muscles are working hard during the course of the day as the head is continually turning from the keyboard in front of the user to the screen in the corner. Keyboard work obviously involves repetitive work for the hands but the repetitive swinging of the head from side to side between screen and keyboard should not be overlooked. Some users assume that the corner of the desk is the 'right' place for the screen simply because that is 'where it has always been'.

The depth of the desk should be sufficient to accommodate the screen and keyboard in a straight line, directly in front of the user and at a comfortable viewing distance. The screen should not be too close to the user – this can cause headaches and eye fatigue. As a general guide, the screen should be approximately an outstretched fingertip distance away when the user is seated comfortably.

The width of the desk should be sufficient to accommodate all work undertaken (not just the DSE tasks). Therefore, if the tasks involve a lot of papers and reference materials, the desk should be able to accommodate this without making the user adopt a cramped working position. It may be necessary to provide two distinct work areas forming an L-shaped arrangement. The use of the old style 'typewriter' return tables may not be appropriate, if used as part of this arrangement. These tables are typically lower than standard desks (they were intended to be used in conjunction with more substantial typewriters and thicker keyboards). If these lower desks are used for anything other than typewriter work, the user will spend much of the time leaning down towards them which may result in backache or possible upper limb discomfort, depending on the position of the wrists and arms.

Workstation providers should be aware that desk design seems, at times, to be one step behind advances in computer design. So although most current desking can accommodate the equipment now in use in most offices, screens are becoming increasingly large and are out-stripping the surface area provided by existing desks.

Drawers

The inclusion of drawers on the desk's undersurface reduces the user's freedom of movement and dictates where they should sit. As the user probably uses the whole of the desk surface, they may find that they have to stretch to reach to the outer areas of the desk as the drawers will prevent them from moving in that direction. The effects of repeatedly reaching to an object, such as a telephone located in the far corner, should not be overlooked. It is very easy to ignore the fact that other activities, on the periphery of the actual keyboard task itself, can affect the user's comfort and well-being.

If desks are to be replaced, consideration should be given to providing users with mobile pedestal drawers. However, these drawers should not, of course, be pushed under the desk out of the way. The pedestal set should be located on the outside of the desk so that the user has freedom to move along the desk length. Another benefit of locating drawers in this position is that it creates an additional surface on which to place items not used on a regular basis and which take up unnecessary space on the desk.

A further problem associated with desk drawers is the effort required to open and close them. Although superficially this may appear to be an insignificant concern, when the effort of using an old and damaged or heavily laden drawer is added to a high-pressure keyboard task, the overall level of effort required to carry out the complete range of tasks is increased. It is easy to focus on the keyboarding activity itself and overlook the other activities which will also have an effect on fatiguing rate. Drawers should be maintained like any other piece of equipment and users should limit the number of items they store in them.

Cable management

Cables should be managed in a safe and efficient manner. It is not unusual to walk around offices and find cables strewn over the floor and under desks, with people using them as footrests. This poses not only safety problems but also task-related ones. It is not uncommon for trips over cabling to result in disconnection of power supplies and consequent loss of screenwork.

The main problem stems from the fact that many cable management systems are not easy to use in the working office. It may appear to be a relatively straightforward process to place cables in a cable trough or through cable wells, or attach them to retaining mounts but trying to change cabling after the initial set-up phase can be extremely difficult. Cable management has to be easy to use and re-use once the desk is functional.

Improvements can be made in the short term if a desk does not have any form of built-in cable management. Cable ties, similar to those used to tie up garden plants, can be used to pull cables into tidy bundles. They are easy to use and to remove once they are no longer required or adjustments need to be made.

Desk edges

Ideally, the user should 'hover' over the keyboard when using the keys. However, users are likely to adopt various keying techniques which include leaning heavily on the desk edge. Many desks, particularly older, traditional-style office desks, have been designed with sharp or 90° angle edges. If users spend long periods at a desk, using the desk edge to support their wrists or forearms during the keying operation, or in between bouts of keying, the sharp edge will protrude into the soft tissue and restrict the blood flow into the hands. This has been associated with the development of ULDs. It would be advisable to provide desks with rounded or 'bull-nosed' edges which will reduce the risk.

Some desks have been given a rounded edge by adding a contoured aluminium strip. Although this reduces the possibility of restricting the blood flow via skin compression, it introduces another problem. It is possible that a degree of heat exchange will occur between the aluminium strip and the skin of the arm as the user leans on the edge. Reductions in skin temperature have also been associated with the development of ULDs.

Undersurface

It is rare for a purchaser selecting a desk to look *underneath* it and it is usually *after* the desk has been introduced into the office that flaws in the design of its undersurface become apparent.

Some desks have a thick frame which runs across the front of the leading edge. As a result, the user may bang their knees on the frame as they pull themselves into position at the desk. Ideally, desks with frames which are directed around and away from the seated user should be selected.

Many desks have 'modesty panels' (originally designed to provide a degree of modesty to female users wearing skirts while working). Given factors such as the increasing number of male users and the use of desks in back-to-back arrangements or in conjunction with screens or against walls, their necessity may not be as universal as it once was. Most modesty panels are designed so that they do not extend all the way to the floor but leave a gap which is just large enough to put the feet under. When the user tries to cross their legs, the panel is in the way of their shins. In addition, many panels are not located at the furthest point away from the user but are set in from the far edge of the desk. This results in a reduction in space under the desk.

The space available under the desk is often reduced by its use as an additional storage area for bags, boxes of materials, files and so on. Sufficient storage facilities should be provided and users should be encouraged to store items as appropriate in lockers, cupboards or filing cabinets.

Footrests

Lack of provision of a footrest – or inappropriate provision of a footrest – is common. As a general rule, if a user's feet do not rest firmly on the floor when using the desk and appropriately adjusted chair, they need a footrest. A footrest should not be given to anyone whose feet naturally rest firmly on the floor (in such a situation, raising the user's feet will result in their lower leg being elevated to a point where their knees are higher than their hips, which may eventually lead to pressure points in their buttocks).

A universal fixed height footrest is unlikely to be adequate. Footrests should be adjustable in height to accommodate different leg lengths. If budgetary constraints are not an issue, gas-lift footrests (which adjust in the same way as a height-adjustable chair) could be provided. A tilt facility should also be considered.

The footrest should have as large a surface area as possible so that the user can move freely when working at the desk. It should be covered in a non-slip matting.

Although a footrest is only designed to be used to support the feet, users may stand on them to open windows, reach to the top of cupboards and so on. It is recommended that they should be constructed of a material that will withstand general misuse (or abuse).

Document holders

It is quite common to observe DSE users inputting information from documents located on the desk surface alongside their keyboard. As a result, they spend much of their time repeatedly looking down at the document and then up at the screen. As is the case with continually moving the head when looking at poorly located screens, the user is again relying on muscles to move the head up and down, and this will be fatiguing. In addition, each time the user looks from the screen to the document, the eyes have to refocus, thus increasing their workload.

The provision and use of a document holder should eliminate the need to move the head and re-focus the eyes. The aim should be to locate the documents so that they are at the same height and distance from the user as the screen. If this is the case, the head does not have to move. Only the eyes will have to glance from side to side, and this is without the need to re-focus continually. To allow the document holder to be used in this manner it should be height-adjustable, or should be capable of being attached to the side of the screen in the appropriate position.

Other 'accessories'

Other items of workstation equipment include wrist rests, VDU arms, mouse mats and screen filters:

- **Wrist rests** are normally about the same length as the leading edge of the keyboard and are located on the desk surface in front of it. They are usually made of a soft material such as fabric or rubber and are particularly useful if an older, thicker, keyboard is in use as they level out the drop from the top of the keyboard to the desk surface. Users should not locate their wrists on these rests and key from that position. To do so would result in them stretching their fingers to reach the keys and this could lead to upper limb problems. Wrist rests should only be used to support the arms in between bouts of keying.
- **VDU arms** have been designed so that a metal plate, large enough to support the screen, is attached to an articulating arm which fastens onto the rear edge of the desk. The arm allows the screen to be pulled into position in front of the user at a height and distance which suits them. Most VDU arms also have a rack attached to the front which can be pulled out to support the keyboard when it is not in use. Once the user has finished with the screen and keyboard (eg when moving on to paperwork), the screen and keyboard can be pushed out of the way, freeing up the desk space. This is particularly useful on desks with a reduced surface area such as the older traditional-style office desks.
- **Mouse mats** are designed to assist in the movement of a mouse, minimising the extent to which it needs to be moved. Foam-based mats are more effective than those with slick surfaces.
- **Screen filters** are intended to control the extent of glare and reflection appearing on the screen and are becoming more common. However, they do not always achieve their aim, and in fact can cause other complications. Sometimes they make fingerprints or cleaning swipes more visible, which themselves become visual irritants. Many users do not realise that they have to clean both the screen filter and the VDU screen at regular intervals. As a result, dirt builds up on the surfaces which makes it more difficult to read the screen. In both cases there is an increase in the workload for the eyes.

Layout

The task of ensuring that the DSE user is as comfortable as possible does not, of course, conclude with the selection of appropriate workstation furniture. Consideration should be given to the layout of the desk surface and the layout of the whole workstation in relation to the remainder of the office.

Worksurface layout

To minimise possible discomfort, it is recommended that users adopt an upright forward-facing posture. The keyboard should be directly in front of them with a space between the keyboard and desk edge which is large enough to rest the hands in between bouts of keying (without this, the arms will not have an opportunity to rest throughout the course of the day).

The screen should be positioned so that it is directly behind the keyboard at a distance which allows the user to view the display comfortably. The user should not have to squint at the text or lean towards or away from it. The only exception to this positioning of the screen is in the case of copy typists who do not have to refer to their screen regularly but spend the majority of their time looking at their documents. In this case, the screen can be moved to the side and the documents placed on a document holder behind the keyboard.

To ensure a reasonably upright sitting position, the screen should be set at a height where the user can read the text without having to raise or lower their head excessively. Given the variation in the dimensions of the user population, it is important that there is a degree of flexibility in the height at which the screen is set. In the short term, if a screen is too low it would be acceptable to prop it up on books or reams of paper until a suitable platform is available. Documents in use should also be moved so that they remain alongside the screen.

Any other items used on a regular basis, such as telephones or reference folders, should be located within easy reach. The user should not have to stretch repeatedly to the furthest points of the desk to pick up objects – particularly heavy items. Items not used regularly can be located on the more extreme areas of the desk surface. If telephone answering is a regular part of the user's job, it may be advisable to provide them with a headset in place of the standard handset.

Workstation layout

How the workstation is positioned within a room may influence the degree of comfort a computer user will experience when working. Positioning the desk so that it is directly in front of a window is a common mistake. The user will then have two contrasting light sources: the outside light which will appear bright; and the darker screen. Although they may think that they are attending only to the information on the screen when working, their eyes will be registering, and dealing with, the two contrasting light sources within their visual field. As a result, the eyes' workload will be increased – often the source of eye fatigue.

Sitting with the back to the window also causes problems. As the light streams in over the user's shoulder, glare will appear on their screen. Again, although apparently concentrating on the screen display, the user's eyes will also be registering the existence of other information. The brain will filter out the redundant information and the user will be unaware that this process has happened, particularly if they are concentrating intensely on their work. However, the extra workload for the eyes will increase the likelihood of eye fatigue.

Ideally, DSE workstations should be positioned so that the screen is sideways on to the window. This is almost impossible in offices where space is at a premium and many people have to be fitted into a small space, or where the office has two or three walls of windows. Blinds or curtains may control problems but they will not necessarily eliminate them. In addition, in an open plan office, users affected by the light may, under pressure from their colleagues, leave the blinds or curtains open and suffer the consequences. Screen filters may be appropriate in these situations.

The workstation's position in relation to overhead artificial lighting is important. If a user is positioned so there is a light directly in front of them, they may experience glare as the light shines in their eyes. If the light is behind them they may see glare or 'hot spots' on their screen. Ideally, they should be located between luminaires. If screen glare is present, the source can easily be located by holding a small mirror over that part of the screen (the mirror will reflect the source of the glare).

IN SUMMARY

- The selection of workstation equipment should be based on the needs of the user population and task demands.
- Current regulations and standards should be used as a guide to determine the initial suitability of workstation equipment.

Seating
- The adjustments should be easy to reach and use.
- Users should be given thorough instruction in the use of the chair and the postures they should adopt.
- Ideally, the backrest should have independent height adjustment.
- The lumbar support of the backrest should be adjusted to the point where it supports the curve of the lower back.
- Armrests should not prevent the user from moving close to the worksurface – if this is the case they should be removed.
- If chairs are provided with armrests, they should be adjustable for height and distance apart.
- Chairs must have a five star base for stability.

Desks
- Either fixed height or adjustable height desks are suitable as long as every user is comfortable when using them. Taller users may require different desks.
- The surface area of the desk should be sufficiently large for the screen and keyboard to be positioned in a straight line directly in front of the user. In the case of copy typists, the screen can be moved to the side and the documents located behind the keyboard.
- The screen should be set at a height and distance from the user which allows for comfortable viewing (both height and distance should be dictated by the user).

- There should be sufficient space in front of the keyboard to rest the wrists on the desk surface when not keying.
- The provision of pedestal drawers should be considered (fixed undersurface drawers restrict freedom of movement). Pedestal drawers should not be positioned under the desk surface.
- Any attached drawers should be maintained properly.
- Users should limit the number of items stored in drawers.
- Desks should have easy-to-use, and re-use, cable management systems.
- Desk edges should be rounded.
- The undersurface of the desk should not cause discomfort to the user in the form of intrusive frames or unsuitably designed modesty panels.
- Users should not store items such as bags and files under the desk. Appropriate storage facilities should be provided.

Footrests

- Footrests should only be provided for those users whose feet do not rest firmly on the floor once their chair has been adjusted for height.
- Ideally, the footrest should be adjustable for height and tilt.
- The footrest should have as large a surface area as possible and be covered with non-slip matting.
- Footrests should be constructed from a material which will withstand the full weight of the user.

Document holders

- Document holders should be provided for users who input information from documents.
- Document holders should be adjustable for height or be capable of being attached to the screen in the appropriate position. They should be located at the same height and distance as the screen.

Other 'accessories'

- Wrist rests should not be used as a base to work from.
- VDU arms will free up desk space.
- Appropriate mouse mats will reduce the degree of hand and arm movement required to move the mouse.
- Screen filters should be selected carefully due to the additional problems they may introduce. They should only be used as a last resort if reflections cannot be controlled by other means (eg blinds).
- Screen filters, and the screens themselves, should be cleaned at regular intervals.

Layout

- Items used regularly should be located close to the user on the worksurface – infrequently used items can be placed at the more distant points.
- The desk should be positioned so that the screen and user are sideways on to windows.
- The desk should be positioned so that the user is between overhead light sources.

DISPLAY SCREEN EQUIPMENT WORK

Workstation use

The seated posture

It is important that the user adopts appropriate working postures and uses workstation equipment properly. The most comfortable position to adopt when using a keyboard is one where the user's elbows are level with the home row (the middle line of characters, starting from the left with 'ASDF'). To achieve this position the user should first adjust the height of the chair. They should aim to adopt a posture where the upper arm hangs naturally by the side of the body with a 90° angle at the elbow and with the forearms parallel with the floor. Such a position will ensure that the user maintains a straight line from the elbow, along the forearm and through the wrist (see illustration below). By ensuring that the wrists are straight at all times when using the keyboard, the user will be less likely to develop a ULD.

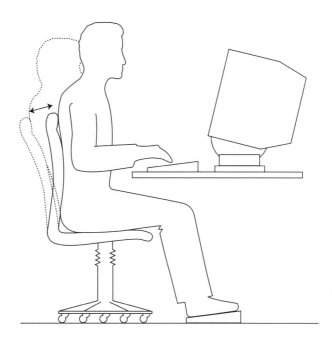

The seated posture

Reproduced from 'Guidance on Health and Safety (Display Screen Equipment) Regulations' (HSE 1992)

Once the chair height has been adjusted, attention should be paid to the feet and the thighs. If the feet are not resting firmly on the floor, or if the front edge of the seat is pressing into the backs of the thighs, the user should be provided with a footrest.

The backrest should be altered for inclination first, then for height (the inclination setting will influence the final location of the lumbar support). The user should aim to sit upright but in a relaxed position. The seated position should not be completely rigid. The user should feel free to alter the setting of the chair as and when they wish, as long as it allows them to adopt a similarly supported and appropriate posture.

Training

Lack of training will undermine the success of any workstation furniture. It is not uncommon for users to misuse their equipment. If users are not given proper training in the use of the adjustments in a chair, the layout of their desk and the postures they should adopt, it is unlikely that they will remain comfortable throughout the course of a working day. Users should be instructed in the full use of their workstation and encouraged, by appropriate styles of supervision, to maintain correct sitting positions.

Trials

The suitability of any workstation item can only be determined by using it in the normal working situation. Therefore, before any final purchasing decisions are made, it is advisable to obtain samples of the intended products in advance for use on a trial basis. These trials should be structured along similar lines to those discussed in the chapter entitled 'The design process'.

Maintenance

All workstation equipment is subject to natural deterioration over time as a result of use or abuse. This should be controlled as far as possible through proper maintenance programmes, and users should be encouraged to report malfunctioning or damaged items as soon as possible.

IN SUMMARY

- The user should adopt a working posture such that the elbow is level with the home row of the keyboard, there is a 90° angle at the elbow, the forearm is parallel with the floor, and a straight line runs through the wrist.
- Users should be trained in the use of their equipment and the appropriate working postures to adopt.
- Workstation furniture should only be selected having been used on a trial basis under normal working conditions.
- All workstations should be maintained at acceptable levels.

Environmental issues

Lighting

If users are to complete their work as efficiently, safely and comfortably as possible, they should be provided with lighting which is appropriate to the workplace and its activities. Poor lighting makes the visual system work harder and may lead to visual fatigue. Common symptoms of visual fatigue include:

- irritation and inflammation of the eyes and eyelids;
- blurred or double vision; and
- headaches, fatigue and nausea.

Poor lighting can also lead users to adopt poor postures in order to view their screen or documents more easily. Such postures will increase the fatiguing rate of the body.

Lighting levels

Illuminance is the amount of light falling onto a surface and this affects an individual's ability to see. Lux is the measure of light falling on a given surface. The BSI suggests that the lighting system should have sufficient flexibility to match the needs of users of display screens and passive media (BSI 2000). Generally, the finer the detail involved in the work, the higher the level of illuminance required. High levels of illuminance, such as 700 lux, will make the characters on a screen less easy to see but will make the task of reading documents easier. Local lighting may be a more appropriate means of highlighting source documents without increasing the illuminance levels beyond those which are suitable for DSE work. Local lighting should be capable of being controlled easily by the person for whom it has been provided.

Glare is a common complaint and is produced by sharp contrasts between different areas in the field of view. Glare can be minimised by:
- ensuring that light fittings are shielded so that the source of light is not visible from the working position (shades or shields may be fitted – up-lighters may be appropriate);
- providing curtains or blinds so that the amount and direction of light through windows can be controlled;
- arranging the room so that the effects of natural and artificial lighting are controlled;
- arranging the equipment on the worksurface so that the effects of natural and artificial lighting are controlled and so that best use is made of local lighting; and
- arranging the equipment so that the screen is at right angles with the window and between rows of luminaires.

Indoor climate

Temperature and humidity

Extremes of temperature may be physically hazardous. High temperatures (particularly if combined with a heavy physical workload, high humidity or low air speed) may lead to dehydration, exhaustion and collapse, abnormalities of cardiac function and various other problems. Between the extremes of temperature and humidity is a relatively narrow range of environmental conditions which may be considered comfortable. Fairly minor deviations from comfort may be subjectively stressful and lead to impaired performance.

The BSI suggests (BSI 2000) that a temperature range of 20–26°C should be acceptable for sedentary work. But there are significant individual differences in this respect, and no single temperature is likely to suit everybody. In a well-controlled indoor environment, the best possible compromise temperature is unlikely to be considered exactly right by more than about 60 per cent of people (and will often be as low as 40 per cent). In general, draughts and thermal gradients (eg between head and feet) should be minimised. For jobs which involve high degrees of seated work, the difference between the temperature at floor level and that at head level should not exceed 5°C. The heat build-up in areas around equipment should not exceed 3°C above the ambient level. Relative humidity of 40–50 per cent will be perceived as being the most comfortable.

The rate of air movement should be less than 0.25 m/s. Where any components of equipment incorporate a cooling fan, the air expelled by the fan should not cause discomfort to users.

Ventilation

If a room has people in it, the air deteriorates in various ways, changing its character due to:

- release of odours;
- formation of water vapour;
- release of heat;
- production of carbon dioxide; and
- air pollution either entering from outside or generated by activities within the room.

The first four arise mainly from the human body itself. The last, air pollution, depends on the situation of the building, what activities are carried out, and whether people smoke. The long term health risks associated with smoking are well known but on a day-to-day basis alone it is likely to cause annoyance as well as eye and throat irritations.

A deterioration in air quality, the situation of the building and the window area are factors to consider in deciding whether it is necessary to have any form of forced ventilation or air conditioning. If it is impossible to open windows for a particular reason, then artificial ventilation is essential.

Noise

Noise is a potential threat to hearing at levels in excess of 85 dB(A). At much lower levels, noise may be a source of stress or may interfere with communication. The individual's subjective response to noise is determined more by its nature and context than its intensity. Thus, intermittent noise and noise with information content, such as background conversations, is much more annoying than continuous, unstructured noise of a similar intensity (eg machine noise).

High concentration work (eg VDU work) and jobs which require high degrees of speech communication are noise-sensitive and even if the noise level is relatively low it can still be disturbing. It is recommended that the noise levels at a VDU workstation should not exceed 55 dB(A) for tasks requiring a high degree of concentration (BSI 2000).

Noisy equipment, such as printers, should stand on a surface which absorbs the sound. If the noise cannot be controlled in this manner then it may be necessary to use acoustic hoods or screens. If noisy equipment will disturb the user it should be moved to another area (if possible).

Health concerns

Health concerns have risen with the increase in computer use. Backache, eye fatigue and ULDs are the more common complaints which keyboard users are concerned about. As discussed above, the provision of an inappropriate workplace and/or the misuse of the workstation are frequently behind many current health problems.

Upper limb disorders (ULDs)

Although a separate chapter is devoted to the subject of ULDs, this section will briefly discuss the issue of keyboard work and its relationship with these disorders.

If a user reports symptoms of a ULD (such as pain, tingling, 'pins and needles') and an investigation is undertaken to determine whether the work is the causative factor, it is not uncommon to find that posture, speed of operation, style of keyboard use and rest breaks are implicated. More recently, the use of a mouse is being added to the list of possible contributory factors.

Posture

The posture adopted by the upper limbs when using a keyboard will greatly influence the likelihood of a user developing a ULD. As stated previously, the aim should be to adopt a posture where the elbow is level with the home row of the keyboard, there is a 90° angle at the elbow, and a straight line runs across the forearm, through the wrist and into the hand.

Because users often neglect to alter their seated position using the adjustments in the chair, and neglect to arrange the layout of their worksurface appropriately, many work with bent wrists, raised shoulders and elbows sticking out from the side of the body. If such postures are adopted and maintained over an extended period, or used repetitively, it is possible that a user could develop a ULD. Posture training has an important part to play in controlling the development of ULDs.

Speed of operation

Today's modern style keyboard allows users to key at faster rates than ever before. And, although it has been recognised that the speed of operation will influence the development of ULDs, many managers or supervisors take advantage of this high speed facility and expect even more work to be carried out within a given timescale. At the top end of the 'at risk' scale are data-entry operators who can key at speeds in excess of 12,000 key depressions an hour (some reach speeds of up to 20,000 an hour). Encouraging users to work to the maximum capacity of their equipment should not be considered acceptable. All keyboard users should be able to work at speeds judged to be comfortable by them.

Style of keyboard use

Many of today's keyboard users have never had any form of training in keyboard skills. A user does not need to be a trained typist to use a keyboard. When an individual joins an organisation and is required to use a keyboard, they are typically shown how to use the software but not the keyboard as such. As a result, there is a population of keyboard users who work in a manner which could contribute to the development of a ULD.

Untrained keyboard users seldom use all their fingers to depress the keys, and will commonly use only one or two fingers of each hand. They do not therefore 'share the load' evenly across the hand, which forces the digits in use to meet all the demands of the keying task. In the case of data-entry operators, their workload may be completed by only one or two fingers on a single hand. It is not surprising that sometimes the limb in use becomes overloaded, presenting with symptoms of a ULD.

Without training, few keyboard users will use 'touch-typing' when inputting information. Given that today's keyboards are 'touch sensitive' and each key requires the slightest of pressure, touch-typing is the appropriate mode of operation. However, many untrained users punch at their keyboards using high levels of force. Keyboards have not been designed to cope with this form of treatment. Each time a slimline keyboard is struck with force, the effect will be similar to hitting the top of a desk with the finger. On impact, a shockwave will pass across the hand and through the wrist. Over time, such repeated shocks could undermine the functioning of the limb, leading to the development of a ULD. Again, training could have an important part to play in the control of ULDs in this situation.

Some of the problems associated with keyboard use have been addressed by designing a new style of keyboard. These keyboards are contoured in such a way as to match the natural shape of the hand when in a 'semi-resting' position. As a result of operating the keys in this more natural position, the hand will be less likely to experience the levels of fatigue associated with use of the standard keyboard design. The use of these keyboards will necessitate additional training, even for experienced keyboard users.

Breaks

All computer users, irrespective of the task they are employed to complete, should take regular breaks in activity. However, many offices have a 'flexible' arrangement where individuals can take breaks at intervals which suit them and the work they are completing. As a result of this 'flexibility' many keyboard users will continue to work, missing opportunities to take breaks, so that they can complete their tasks on time. Although it may be perceived as inefficient to stop work at regular intervals, it is actually a much more effective way of working. The rest breaks prevent the onset of fatigue and subsequent deterioration in performance which usually develops over the course of a day. The duration and regularity of breaks is discussed in more detail in the chapter on 'Job design and work organisation'.

Mouse use

Complaints resulting from the use of a mouse have increased steadily. It is not surprising that users have reported symptoms of pain given the circumstances surrounding mouse use:
- Training in mouse operation is not common and usage is often sustained for long periods.
- Many people work without a mouse mat which may result in increased movement of the hand and arm.
- The mouse is typically located on the outer edge of the keyboard and the user has to extend their arm and reach towards it.

- Few people will have been advised to maintain minimal pressure when gripping the mouse. Therefore, they may spend long periods gripping it firmly. Forceful repetitive gripping can promote the onset of ULDs.

Users should be trained in the use of the mouse, as with any other piece of equipment. They should be advised to:

- use a suitable mouse mat;
- keep the mouse as close as possible;
- maintain minimal pressure during use;
- release the mouse when it is not in use and return the hand to a more natural position; and
- move the keyboard slightly to the side when it is not in use (eg during an editing exercise) so that the mouse can be brought closer.

Special care should be taken if users are operating a mouse which is attached to a laptop computer. Often the mouse will be attached in such an orientation that it forces extreme deviation of the wrist, thereby making it unsuitable for use over extended periods. Some laptops are directed by use of a button which is manipulated by the tip of a finger. Again, prolonged use could be problematic. In such circumstances, it may be advisable to use a 'standard' mouse with the laptop.

Other health concerns

Pregnancy

There is a great deal of concern regarding the hazards to unborn children as a result of their mothers' DSE work. The HSE has stated (HSE 1992) that when the scientific research is taken as a whole, the results do not show any link between miscarriages or birth defects and working with DSE. Such research is continuing.

Irrespective of the current consensus of scientific opinion regarding pregnancy and DSE work, women who are pregnant, or who are planning to start a family, should be given an opportunity to discuss their concerns with an appropriate member of the organisation, and their concerns should be dealt with sensitively and sympathetically.

Skin irritations

Skin rashes, facial dermatitis and dry skin are sometimes reported by DSE users. Generally, these conditions are not considered to be related directly to DSE use but occur because of the environment in which the users work. Examples of such causes include static electricity and low relative humidity.

Epilepsy

The HSE has advised (HSE 1992) that individuals who suffer from epilepsy, including photosensitive epilepsy, can work safely with DSE. However, epilepsy sufferers should be able to discuss concerns relating to their condition with an appropriate member of the organisation.

Visual fatigue

Visual fatigue results from the effort of trying to see, and does not necessarily mean that there is a deterioration in the quality of the eyes. Symptoms include redness and irritation, blurred vision and headaches. These can result from working with DSE for long, uninterrupted periods of time, particularly if the work requires high levels of concentration. Concentration often results in a user staring at the screen for long periods. Staring results in a reduction of blinking which normally ensures that the eyes remain moist and free from dust and dirt. Poor lighting levels, flickering lighting or a flickering screen, glare and reflections, and poor screen or source document legibility can also cause eye fatigue. Some users have reported reductions in headaches as a result of using 'daylight' bulbs.

Stress

Stress is becoming a more prominent issue in the workplace. It is considered that monotonous work, high perceived workload, time pressure, low control and minimal social support all contribute to feelings of stress (Moon & Sauter 1996). Such conditions are commonly associated with keyboard operations. It could, therefore, be considered that the design of keyboard tasks may influence the level of stress a user will experience when at work.

An association between stress and the development of symptoms of a ULD has now been established. It is considered that mental tension (or stress) can have a 'knock-on' effect in terms of physical tension. For example, when driving in heavy rain on a motorway, a driver may become anxious when overtaking a large vehicle creating a wash of dirty water in front of them. Once the driver has safely passed the vehicle and pulled back into the inside lane they may realise that the muscles in their shoulders and neck had become very tense, only relaxing after the danger had passed. It is now considered that such increases in physical tension, resulting from raised stress levels, can increase an individual's susceptibility to developing a ULD.

Exercise

Some organisations are introducing exercise programmes into daily work routines as a means of combating the effects of users sitting for long periods of time in the same posture. It may not be that the exercise itself is of benefit but rather that the benefits stem from the fact that the user is given an opportunity to change position and to stop work for a period of time. Either way, exercise should be given serious consideration, particularly in environments that have been identified as high risk. Exercise programmes should be developed by appropriately qualified people and the emphasis should be on gentle exercise.

IN SUMMARY

Environmental issues
- Lighting levels should be appropriate to the task being performed.
- Blinds or curtains should be fitted to windows.

- Screens should, where possible, be positioned at right angles to windows and between rows of lights.
- Temperature levels should be appropriate to the tasks being performed.
- Draughts and thermal gradients should be minimised.
- The air quality, situation of the building and window area should determine the need for ventilation or air conditioning.
- Noise should be controlled as far as possible in high concentration areas.

Health concerns
- Users should adopt appropriate working postures to reduce the likelihood of developing a ULD.
- Keying speeds should be controlled.
- Training in keyboard skills should be considered as a means of reducing the likelihood of injury occurring.
- Users should have regular rest breaks and/or changes in activity.
- Users should be given instruction in the appropriate use of a mouse.
- Users who are pregnant, suffer from epilepsy or other health concerns such as skin irritations should have the opportunity to discuss their condition with an appropriate individual within the organisation.

REFERENCES AND FURTHER READING

Beardon C and Whitehouse D, 1993, *Computers and society*, Intellect, Oxford.

British Standards Institution, 1999, *Ergonomic requirements for office work with visual display terminals (VDTs) – Part 5: Workstation layout and postural requirements*, BS EN ISO 9241-5, BSI, London.

British Standards Institution, 2000, *Ergonomic requirements for office work with visual display terminals (VDTs) – Part 6: Guidance on the work environment*, BS EN ISO 9241-6, BSI, London.

Chartered Institution of Building Services Engineers, 1993, *Lighting for offices*, CIBSE, London.

Chartered Institution of Building Services Engineers, 1994, *Code for interior lighting*, CIBSE, London.

Grandjean E, 1994, *Ergonomics in computerized offices*, Taylor & Francis, London.

Health and Safety Executive, 1992, *Guidance on Health and Safety (Display Screen Equipment) Regulations*, L26, HSE Books, Sudbury.

Health and Safety Executive, 1994, *VDUs: an easy guide to the Regulations*, HSG90, HSE Books, Sudbury.

Health and Safety Executive, 1995, *How to deal with sick building syndrome: guidance for employers, building owners and building managers*, HS(G)132, HSE Books, Sudbury.

Health and Safety Executive, 1997, *Seating at work*, HS(G)57, HSE Books, Sudbury.

Kerr J, Griffiths A and Cox T (eds), 1996, *Workplace health, employee fitness and exercise*, Taylor & Francis, London.

Kirby M A R, Dix A J and Finlay J E (eds), 1995, *People and computers*, Cambridge University Press, Cambridge.

Leuder R and Noro K, 1994, *Hard facts about soft machines: the ergonomics of seating*, Taylor & Francis, London.

Moon S D and Sauter S L (eds), 1996, *Beyond biomechanics: psychosocial aspects of musculoskeletal disorders in office work*, Taylor & Francis, London.

Morris A and Dyer H, 1998, *Human aspects of library automation* (2nd edition), Gower, Aldershot.

Sauter S L, Dainoff M and Smith M, 1990, *Promoting health and productivity in the computerized office*, Taylor & Francis, London.

HAND TOOL DESIGN AND USE

The use of hand tools can be traced back to the Stone Age. Their purpose was to extend the use of the hand, allowing it to complete tasks which it would not be capable of normally. Tools allowed the hand to centralise and deliver power resulting in the slicing, cutting, smashing, piercing or scraping of an object. That purpose and capability remains unchanged today – but, of course, today's tools are made from different materials and many are power-driven. Unfortunately, the modern adaptation of the basic tool design has resulted in an increase in the potential for injuries. But that is not to say that a simple tool, such as a hammer, is without its problems.

Injuries occur as a result of a tool being designed solely from a 'function' perspective which ignores the effect its form can have on the user. Obviously, meeting the requirements for use is important but this does not have to be achieved at the expense of the user. If a poorly designed tool is used day in, day out, this could result in decreased productivity as well as injury.

In some cases, users are not provided with any tools and find that they have to use their own hands. For example, it is quite common for workers involved in assembly operations to use their hands like mallets when pushing components into position. This practice can have a destructive effect on the limbs and should not be allowed to continue.

The hand

The starting point when designing a tool should be an examination of the hand as this is the means by which the user will hold and operate the tool.

Gripping

The hand is a versatile, flexible tool in itself and possesses a neuromuscular power unsurpassed by any other part of the body. It can be used for extremes of function ranging from the primitive grabbing of an object, such as a spanner, to the delicate manipulations required during a fine motor task, such as making delicate pieces of jewellery. The hand's use of such power grasps or precision grips will be determined by the characteristics of the object being gripped and the task demands.

To understand the concept of the power grip, imagine the hand making a fist where the fingers are wrapped around one side of an object and the thumb around the other (eg when gripping the handle of a hammer). There are three separate categories of the power grip, distinguished by the direction of the force:

- parallel to the forearm (eg as when using a plane or saw);
- at an angle to the forearm (eg as when using a mallet); and
- rotating about the forearm (eg as when using a screwdriver).

The precision grip is used for precise manipulations which require the object to be held between the fingertips and thumb. Examples of precision grip tasks include the picking of a loose hair from clothing, turning a key in a lock or using a pen. The precision grip can be either internal or external. Holding a pen is an example of an external precision grip because the object is outside the palm. Holding a knife when eating a meal is considered an internal precision grip as the knife handle is held inside the hand.

Often, a combination of both power and precision grasps will be used to complete an operation. For example, when returning the lid to a soft drink bottle a precision grip is used to hold and locate the lid carefully before switching to the power grip for final tightening to ensure that it is secure. It is considered (Putz-Anderson 1988) that the power grip provides the individual with five times the gripping strength of the precision grip, which suggests that the strength requirements of an activity can be reduced if the task can be completed using the more powerful grip. In terms of tool design, if individuals can use a power grip to hold a tool they can use a lower percentage of their overall gripping strength, which will in turn be less fatiguing for their muscles.

Force

The hand can exert the most force when in a neutral, or near neutral, position (ie when a line runs virtually straight from the elbow through the forearm and wrist and into the hand). Once the hand is bent downwards or to either side of the wrist, the grasping power of the hand is reduced. The loss of grip strength following deviation of the wrist is illustrated below.

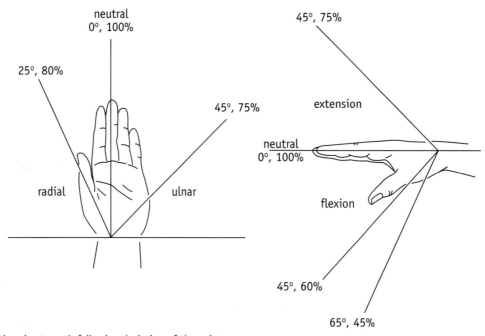

Change in grip strength following deviation of the wrist
Reproduced from 'Cumulative trauma disorders: a manual for musculoskeletal diseases of the upper limbs' (Putz-Anderson 1988)

It would appear therefore, that in order to allow the hand to work most effectively it should be permitted to adopt a neutral posture and this should be taken into consideration during any tool design or selection.

HAND TOOL DESIGN AND USE

Design features

Hand or arm injuries following the use of tools may indicate inappropriate tool design, or improper use. Such injuries can range from something as simple as a callus or blister, to something more severe such as an upper limb disorder. In most instances, injuries could be avoided by ensuring that tools are properly designed, taking both the user and the task into account.

Handle design

One of the most common complaints made by workers is that they have to bend their wrist when using tools. For some reason designers have, up until recently, designed most tools with straight handles and heads. As a result, when the tool user works on or against a flat surface they will usually have to bend their wrist.

The user experiences a loss of grip strength from working with a bent wrist. To overcome this problem, and to control the tool fully, the pressure of the grip may be increased. Increasing the grip pressure will speed up the rate at which the muscles fatigue. Therefore, to avoid the reduction of grip strength and subsequent acceleration of muscle fatigue, maintaining a straight wrist should be a primary design and selection consideration. This can be achieved easily in many work situations by bending the handle (see below).

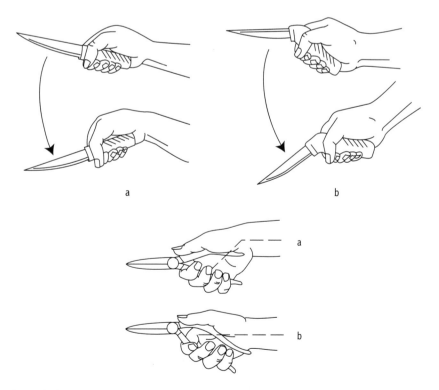

Changing the wrist angle by bending the handle

Reproduced from 'A guide to the ergonomics of manufacturing' (Helander 1995) (top) and 'Cumulative trauma disorders: a manual for musculoskeletal diseases of the upper limbs' (Putz-Anderson 1988) (bottom)

HAND TOOL DESIGN AND USE

In some situations, bending of the wrist can be eliminated by providing a pistol grip tool where the handle is bent by 70–90°. However, this design is only acceptable if used in an appropriate orientation. In general, pistol grip tools should only be used when the tool axis is horizontal. If the tool axis is vertical or the force is applied perpendicularly to the work plane, a straight grip should be used. Examples of wrist posture, tool design and orientation of use are shown below.

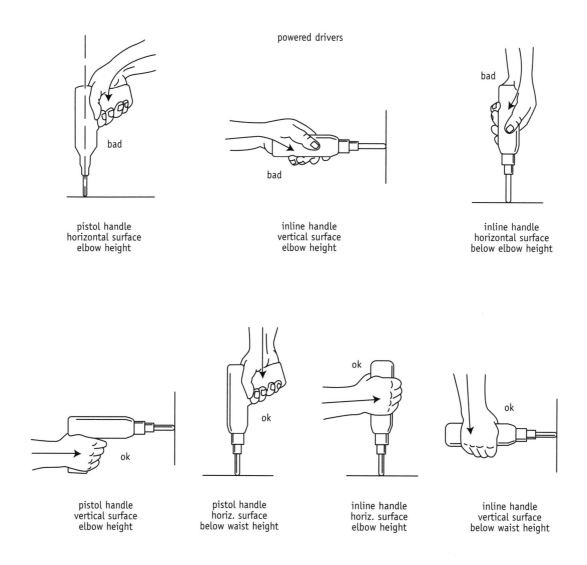

Handle designs, orientations of use and wrist postures

Reproduced from 'Cumulative trauma disorders: a manual for musculoskeletal diseases of the upper limbs' (Putz-Anderson 1988)

Handle construction

If users have to work with a tool which has a smooth, hard handle constructed of a material such as flat metal or plastic, they may experience difficulty in stopping their hand from sliding across the handle as they apply force. To prevent the hand from moving, the user will increase the pressure of the grip – this situation is made worse if their hands are hot and sweaty.

Care should be taken if the handle is designed with flutes or ridges (eg a screwdriver). If these flutes or ridges are too deep or have sharp edges, they may increase the pressure placed on the soft tissue of the hand when the user grips the handle and this may result in discomfort and pain. Soft oval indents on a handle are preferable – they allow improved purchase without causing pressure points.

The use of metals to form handles may result in the handle remaining cold throughout the course of the shift. Exposure to colder temperatures may increase the possibility of a user developing a ULD, therefore the use of metals in handle design should be avoided.

It is recommended that soft, compliant and textured materials are used on handles so that they are easier to grip, although care needs to be taken that these do not damage easily, subsequently presenting sharp edges.

Application of force

If users have to apply downward force when working with a vertically orientated tool, such as a drill fitted with a parallel sleeve, they may find that their hands tend to slip downwards during the operation. In an attempt to prevent this, they will increase the force used to grip the sleeve. The need to use this level of additional force could be controlled by simply fitting a moulded collar to the bottom of the sleeve. This would prevent the hand from moving and would allow the user to apply the downward pressure necessary to operate the tool without the need to increase grip pressure. A similar approach could be taken to the design of handles on tools such as pliers where thumb stops could be added to prevent the hand from slipping. Any such alterations will obviously be dependent on the particular operation being performed.

Contouring

Contouring along the handle may not achieve the intended result of assisting the user to hold it firmly and comfortably. Contouring, particularly finger grooves, should be avoided altogether unless the design is based on anthropometric dimensions and will actually 'fit' the users' hands sufficiently. If the finger grooves have not been based on anthropometric dimensions, users may find that their own fingers do not fall in line naturally with the grooves, necessitating an alteration in their natural gripping position to suit these contours. Deviation from the comfortable or natural gripping position of the hand will result in a reduction in grip strength and an increase in pressure exerted by the user to ensure that they hold the tool firmly.

Handle size

Grip strength required to complete a task will also be influenced by the diameter of the handle. It is not uncommon for a small tool to be provided with an equally small handle. However, reducing the size of the handle makes it more difficult to grip. This is also the case if the handle is too big to fit comfortably into the hand. In both cases, users will have to increase the pressure of their grip to ensure that they maintain a firm hold of the tool and control it during use. It has been recommended that the diameter of a single-handle tool should be approximately 40 mm, which will be suitable for most purposes (HSE 1990).

However, for situations where torque is exerted about the axis of the handle, as in the case of using a screwdriver, it may be advisable to provide a handle with a larger diameter, possibly up to 65 mm. Although there may be a decrease in grip strength as a result of going beyond the optimum handle diameter, this may be offset by the increase in mechanical advantage.

If the handle of a tool is too short it may press into the palm of the hand during use. This is quite common when using tools such as pliers or paint scrapers. The palm is rich in nerves and blood vessels and compression by the handle can lead to reduction of blood flow and nerve damage. Many users working with short-handled tools report feelings of numbness and tingling and this can become more serious with long term use of the tool. The handle should be at least 100 mm in length. To provide users with some freedom in terms of where they place their hand along the handle, a length of 115–120 mm is preferable. The desirable handle length will be influenced by the use of gloves, therefore a further 10 mm should be added to the length if they are worn.

Double-handle tools

A number of tools such as buffers, grinders, drills and sanders may have two handles or the facility to position the second hand on the tool unit. The advantages of such a design are that it shares the demands of the operation out more evenly between the hands and may improve the positioning and control of the tool, particularly for fine, precise or delicate work. However, it is hard to find double-handle tools which have two handles of equivalent size and construction. For example, many angle grinders require the user to grip the main body of the grinder (which houses the motor) while the other hand grips a much thinner plastic handle set at 90°.

Clearly, the general requirements and specifications for handle design covered in this chapter apply equally whether there is one handle or two.

Handedness

Designers typically develop tools for use by right-handed users which causes problems for left-handed users (representing 10 per cent of the workforce). If left-handed users attempt to use their right hand they may find that their non-dominant hand does not have the same level of dexterity or the same strength capacity as their dominant hand. If they use the tool in the left hand they may find that any contoured finger grooves or apertures present do not allow them to grip the handle easily and comfortably (eg scissors). It is therefore important to select tools, whenever possible, which allow for use by either hand. If this is not possible, left-handed users should be provided with tools designed specifically for their use.

Handle span

The span between two handles of a tool will influence the type of grip used and the strength requirement of the task. For example, the operation of nail scissors will result in the user applying a pinch grip to hold the scissors which will reduce the effective grip strength by up to 25 per cent. To control the degree of effort required to close the handles, it is generally recommended that the span should be 50–75 mm. If the tool is used repetitively, as in the case of using scissors to remove threads from clothing, it is recommended that the handles should be fitted with an automatic spring-opener. This will allow the use of the stronger hand-closing muscles and avoid the need for using the weaker hand-opening muscles. Care should be taken with the tension setting on the spring; the user should not have to exert excessive force to close the handles due to the resistance of the spring.

Aperture size

If the hand or fingers have to be inserted into an aperture, such as when using scissors or a saw, consideration should be given to finger and hand clearance. This is particularly important because of the sensitivity of the backs and sides of the fingers. Repetitive use of standard surgical-type scissors for a finishing task, such as the removal of excess material from moulded rubber components, will result in the exertion of pressure on the backs of the fingers each time the blades are opened. In addition, the concentration of pressure around the joints of the fingers, due to the reduced dimensions of the finger apertures on the scissors, will compromise the functioning of the fingers. A finger ring with a diameter of 30 mm will allow for the comfortable insertion, removal and movement of a finger or thumb by most users. If the hand is to be inserted into an aperture, as when using a saw, it is recommended that a rectangle of 110 mm by 45 mm is provided.

Triggers

The inclusion of a trigger on a tool introduces further problems requiring consideration. If the trigger is to be used repetitively, as in a production setting, it should be capable of being activated by more than one finger. This will allow the user to 'share the load' more evenly across the fingers, thereby minimising the possible fatiguing effect on any single digit, while at the same time increasing strength capacity. To allow for activation by more than one finger, a 'trigger strip' of 50 mm should be provided. If activation by one finger only cannot be avoided, the handle should be of a sufficient length to be capable of being gripped by the whole hand when the finger is off the trigger, outside the trigger guard. This will prevent accidental activation of the tool when it is being moved.

If the handle is so large that the user has to reach with the finger(s) to operate the trigger, they will only be able to flex (or bend) the tips of their fingers when applying the necessary force. This can cause pain and discomfort in the hand and fingers, and over time can result in the development of a ULD. The handle size should allow for activation of the trigger by the middle and end sections of the fingers simultaneously.

Triggers which are hard, or which have sharp edges, can also cause compression of the soft tissue of the fingers and may produce feelings of discomfort. The trigger should be covered with a similar material to that used on the main handle and should be rounded to eliminate any sharp edges.

The effort required to depress the trigger is also a consideration. The level of effort judged acceptable will change in line with the frequency of use. Generally, the more frequently a trigger is used, the less effort should be required. However, the degree of pressure required to activate the trigger should be compatible with the avoidance of inadvertent operation. Accidental start-up can obviously be hazardous but increases in static loading, resulting from the user deliberately holding their finger away from the trigger for fear of accidentally depressing it, can also lead to problems.

Static and dynamic work

Static muscle work is required to hold the tool, while dynamic muscle work is required to manoeuvre it into position and use it. Static muscle work requires the muscles to tense for an uninterrupted period of time. It is considered far more fatiguing than dynamic muscle work which allows the muscles to rhythmically contract and relax during the operation. It has been suggested that muscles involved in dynamic work are more resistant to fatigue and possible injury than muscles involved in static work. Therefore, it is essential that the effects of static muscle work are minimised as far as possible. The weight of the tool and the distribution of weight both impact on static work.

Weight

A frequently used tool should weigh as little as possible. It has been suggested that it should be capable of being held in one hand and not weigh more than 0.5 kg. (It is possible to source alternative tool designs which complete a job efficiently while weighing only half as much as their regular counterparts.) If a tool weighs more than 0.5 kg, consideration should be given to using a counter-balance. If the tool is not counter-balanced the user should be encouraged to put the tool down at every appropriate opportunity to allow the hand to recover from the static muscle work required to grip it during use. Even a break of a few seconds will make a difference to the fatiguing rate of the muscles. To encourage users to put their tools down they should be provided with a surface or holster close to the tool's area of use. Alternatively, if the tool's use, shape and weight allow it, straps around the handle, such as those found on ski poles, would allow the user slight relief from the static loading.

There are two exceptions to the rule of minimising tool weight: those tools which need a certain weight to prevent the transmission of vibration to the user's hands and arms; and those which enable equal pressure to be exerted over the surface being addressed (eg as when using a buffer or grinder). If users find that the tool is too light to carry out the job efficiently, they may try to compensate by gripping harder and exerting higher levels of force.

The tool's centre of gravity should be close to the hand (the effort to grip the tool will increase as the centre of gravity moves away from the hand). Locating the handle close to the tool's centre of gravity should also minimise the possibility of the tool slipping out of the hand. However, when determining the location of the handle, consideration should be given to the influence of air hoses and any other attachments which might alter the distribution of weight. To control their influence, the length of any air lines and power cables should be kept to a minimum. The aim should be to eliminate the need for the arm muscles to compensate for inappropriate tool balance.

Air hoses

If users are working with tools with an air hose they should be protected as far as possible from the cold 'blow-back' from the tool. The cold air jet is often directed towards the hand and wrist area and the subsequent reduction in temperature may result in the reduction of blood flow into the hand, which is associated with the development of ULDs. In addition, when the hands become cold, the user will experience a loss of dexterity and grip strength and as a result may have to increase the pressure of their grip. If gloves are used as a means to limit the effects of the cold air, several issues should be given careful consideration:

- The gloves should fit the hand properly so that they do not impair manual dexterity.
- Tight gloves may deceive the user, leading them to believe that they have a firm grip on the tool when they do not, which could result in tool slippage and a possible accident.
- Loose or bulky gloves may become caught up in rotating tool elements.
- Bulky gloves may interfere with tactile feedback, resulting in the user gripping the tool more firmly than is necessary.
- Additional clearance may be required for the gloved fingers and hands.

Vibration

There are concerns that hand-held power tools which transmit vibration to the hand and arm may be responsible for a series of ULDs. Tools which are considered to carry higher levels of risk include chainsaws, grinders, jackhammers, concrete vibrators and pneumatic drills. Vibration is known to cause constriction of the blood vessels in the hand and a subsequent reduction in blood flow, resulting in the white appearance of the fingers and hands commonly associated with the disorder Vibration White Finger. Users who operate vibrating tools may report feelings of pain, numbness or 'pins and needles', all of which affect manual dexterity. In very severe cases, the blood circulation may be permanently impaired and the fingers may appear blackish-blue in colour. In extreme cases, gangrene may develop, although it is very rare for the condition to progress as far as this – action is usually taken on recognition of the early symptoms.

As a consequence of the reduction in manual dexterity indicated above, and the need to hold a vibrating tool steady, the user often grips the tool more firmly which results in an elevation in the fatiguing rate of the muscles and an increase in the transmission of vibration from the tool to the hand.

Vibration with a frequency ranging from about two to 1500 Hertz (cycles per second or Hz) is potentially damaging, and is most hazardous in the range from about five to 20 Hz (HSE 1994a). However, the risk is influenced by the user's daily 'vibration dose', ie the vibration magnitude and exposure time. The severity of the risk is also affected by factors such as the work-rest schedule, how much of the hand is exposed to the vibration, and other influences on blood circulation such as smoking, temperature and heredity.

There are several methods for controlling the amount of vibration transmitted to the user's hands, including:

- The selection of appropriate tools for the tasks.
- The provision of appropriately designed handles which reduce the grip strength required to hold, locate and use the tool.
- The provision of adequate training so that lack of experience does not lead to the use of excessive gripping forces.
- Interruption of tool use with regular breaks.
- The appropriate maintenance of tools.
- The use of anti-vibration mountings and vibration-isolation handles.
- The substitution of the operation with another less stressful one which will achieve the same end result.

Wearing gloves will not necessarily diminish the effects of the vibrating tool as they may impair manual dexterity, resulting in an increase in the force used to grip the tool.

Torque response

It is not uncommon for users to experience upper limb pain caused by the 'kick back' from the recoil of tools such as nut runners. The abrupt snap response of the nut runner indicates that a nut has been driven to the required level of torque, and having achieved this the tool switches itself off immediately. Should the tool fail to switch off, the torque may be relayed to the user's arm which can result in injury.

The sudden stress placed on the hands and arms by the abrupt switching off of the tool should, if possible, be controlled by using variable torque tools. These have a clutch mechanism which will automatically alter the effective torque as the required level of tightening is reached and will shut the tool off gradually, thereby eliminating the sudden snap-off which can prove to be destructive over time.

Multi-functional tools

To eliminate the need for tool users to switch continually between a series of different items, repeatedly picking them up and putting them down, it might be advisable to design or select a multi-functional tool capable of carrying out several operations. However, the advantages of using one tool continually should be weighed off against the benefits of allowing the hand to release one object and then grip an alternative shape or weight.

Domestic and off-the-shelf tools

It is not unusual for domestic tools to be introduced into a work setting, although many are not designed to be used for long periods or in a repetitive manner and are therefore unsuitable for occupational use. For example, a domestic hand drill may wear out quickly and vibrate more than is considered acceptable.

Similarly, 'off-the-shelf' tools are often used in inappropriate environments. As a result, the user may have to compensate with inappropriate limb postures and increased forces. Only appropriately designed work tools should be used in a work environment.

Areas of use

Having given full consideration to the tools themselves, attention should also be focused on their area of use as this will influence the ease with which a tool can be used and a task completed. Ideally, any surfaces or components which are being addressed by the tool should be oriented so that the user can remain in a natural, upright posture. They should not have to deviate their body or limb posture grossly in order to apply the tool.

Access to the application area should be made as easy as possible. Any obstacles between the user and the area of application should be removed so as to minimise unnecessary reaching.

If the tool user has to lean on a surface to steady a hand during work, consideration should be given to padding the surface to increase comfort levels and to reduce the possibility of contact with any sharp edges or cold surfaces.

In conclusion

It is considered that the handle is the easiest piece of the tool to alter, yet it is the part that is usually overlooked during design and redesign processes. A properly designed tool handle should enable full tool control and stability, increase mechanical advantage and reduce the amount of effort required. Poor tool design results in users having to adopt a series of irregular postures which allow them to hold and use the tool. Gross deviations in posture and subsequent increases in required effort result in an elevation of fatiguing rate and the possible development of discomfort, pain or injury. Tool designs should accommodate the user and enable them to work comfortably and safely – while allowing efficient completion of the operation.

IN SUMMARY

Gripping
- At every opportunity, the user should be able to employ a power grasp when holding a tool.
- The user should be able to use the tool when the hand is in the most natural position, with a straight line running from the elbow through the wrist and into the hand.
- To avoid bending the wrist, the tool handle or tool head should be bent.
- Pistol grip tools should only be used in an orientation which permits the wrist to be held straight (otherwise a straight grip tool should be used).

Handle design and construction
- A tool handle should not have a hard or smooth surface which will allow the hand to slip during use.

- When applying downward force with a tool (eg a drill), a moulded collar will prevent the hand from slipping as pressure is applied.
- Handles made from metal which may be cold should be avoided due to the possibility of reducing the temperature of the hand.
- The handle should be covered in a soft, compliant and textured material to make it easier to grip.
- If flutes or ridges are used on the handle they should not be too deep or have sharp edges which might damage the soft tissue of the hand.
- Contouring on handles such as finger grooves should be avoided unless their design has been based on anthropometric data specific to the intended user group.
- A single-handle tool should be approximately 40 mm in diameter but an increase to 65 mm may be acceptable if torque is exerted about the axis of the handle (eg when using a screwdriver).
- The handle should be a minimum of 100 mm in length with a preference for 115–120 mm. A further increase of 10 mm should be made if gloves are worn.

Handedness

- Tools should be designed for operation by both left- and right-handed users. Alternatively, specifically designed tools for left-handed users should be provided as appropriate.

Handle span

- The span between two handles (eg as with a pair of pliers) should be 50–75 mm.
- For repetitive use, an automatic spring-opener should be added to tools with two handles (eg scissors). The spring's tension setting should not demand excessive levels of effort to close the handles.

Aperture size

- Finger apertures in tools such as scissors should have a diameter of 30 mm.
- Hand apertures in tools such as saws should form a rectangle of 110 mm by 45 mm.

Triggers

- If a trigger is used repetitively, it should be designed so that it can be activated by more than one finger.
- A 'trigger strip' of 50 mm is recommended to allow for activation of the trigger by more than one finger.
- If only one finger is used during trigger use, the handle length should be such that it can accommodate all the fingers when the trigger finger is not in use.
- The trigger and handle dimensions should be co-ordinated so that the trigger can be activated by the middle and end sections of the fingers, not just the tips.
- Triggers should not be hard or have sharp edges.
- The effort required to depress the trigger should be regulated and altered in line with the frequency of use.

Static work

- The static effort required to hold and use the tool should be controlled as far as possible.

Weight

- The tool weight should be kept to a minimum (except for those tools which require weight to prevent transmission of vibration and those which enable even pressure to be exerted over a surface).
- Heavy tools should be supported by a counter-balance or placed in a holster when not in use.
- Users should be encouraged to put the tool down when not in operation.
- The centre of gravity of the tool should be kept close to the hand.
- The length of air lines and other attachments should be kept to a minimum to limit the effect on the tool's balance.

Air hoses

- Users should be protected from the cold air blown back from an air tool.

Vibration

- Vibration should be kept to a minimum and users protected from it as far as possible.

Torque response

- The torque response of tools should be controlled to limit the effect on users.

Multi-functional tools

- Where appropriate, multi-functional tools should replace a series of individual tools.

Domestic and 'off-the-shelf' tools

- Domestic and 'off-the-shelf' tools should not be used in an occupational setting unless they are suitable.

Areas of use

- Access to the application area should be made easier by inclining surfaces when necessary, removing obstacles and padding any surfaces which may be leaned on.

REFERENCES AND FURTHER READING

Health and Safety Executive, 1990, *Work-related upper limb disorders: a guide to their prevention*, HSG60(rev), HSE Books, Sudbury.

Health and Safety Executive, 1994a, *Hand arm vibration*, HSG88, HSE Books, Sudbury.

Health and Safety Executive, 1994b, *A pain in your workplace?*, HSG121, HSE Books, Sudbury.

Health and Safety Executive, 1997, *Vibration solutions: practical ways to reduce the risk of hand-arm vibration injury*, HSG170, HSE Books, Sudbury.

Helander M, 1995, *A guide to the ergonomics of manufacturing*, Taylor & Francis, London.

International Labour Office, 1996, *Ergonomic checkpoints*, ILO, Geneva.

Pheasant S, 1996, *Bodyspace – anthropometry, ergonomics and the design of work* (2nd edition), Taylor & Francis, London.

Putz-Anderson V, 1988, *Cumulative trauma disorders: a manual for musculoskeletal diseases of the upper limbs*, Taylor & Francis, London.

JOB DESIGN AND WORK ORGANISATION

The job design process

Providing workers with well-designed workstations, workplaces and equipment is not enough to ensure that they can work comfortably, safely and contentedly. Job and work organisation factors – such as speed of operation, rest breaks and rotation programmes – also need to be carefully designed as they will determine how people relate to each other and with the production system.

There are three ways of looking at the needs of the organisation during the design process:

- **Production system framework** – the organisation is viewed as one flowing process whereby items, such as a product or service, are fed into the system and transformed into an output. To facilitate this framework, the job must be designed to meet the requirement for a high quality output. Therefore, the focus will be on generating efficient operations, maintaining equipment, recording processing information and controlling the quality of materials.
- **Miniature society framework** – this perspective views the organisation as a social institution consisting of individuals who act in response to shared experiences, expectations, rewards, conflicts, prejudices and so on. To facilitate this framework, the job design and work organisation must meet the needs of recruitment, training, co-ordination, communication and other related issues.
- **Individual framework** – this framework acknowledges the importance and contribution of the individual within the organisation. It recognises that each individual has a specific career path in mind and has personal expectations and aspirations. There is also a recognition that the individual's outlook does not necessarily focus solely on the organisation but certainly overlaps with it at intervals. An organisation is more likely to meet its own goals if it is able to design jobs and work systems which allow the individual to satisfy personal needs and achieve personal goals.

There are various desirable features which are important in job design if both the organisation and the individual are to satisfy their needs and achieve their aims:

- The work should be challenging and meaningful.
- The individual should be able to learn 'on the job' (this will necessitate an indication of specific performance criteria in advance and feedback following performance).
- The individual should have some control over decision-making and be permitted to use their judgment and discretion. Achievements should be measured by evaluating objective outcomes.
- There should be social interaction where individuals can call on each other for support, assistance and understanding.
- There should be a recognition of each individual's contribution.
- The individual should be able to equate their work role with their role outside work.
- Individuals should perceive that they have a desirable future and are not stuck in 'dead end' jobs.
- The job should be flexible enough to accommodate individual differences, characteristics and circumstances.

The 'Quality of working life criteria checklist' (see below), linked to the list of desirable features listed above, presents individual needs in a succinct format. It provides a profile of the factors which are of importance to an individual and which influence their levels of satisfaction/dissatisfaction in the workplace.

Physical environment	Safety, health, attractiveness, comfort
Compensation	Pay, benefits
Rights and privileges	Employment security, justice and due process, fair and respectful treatment, participation in decision-making
Job content	Variety of tasks, feedback, challenge, task identity, individual autonomy and self-regulation, opportunity to use skills and capabilities, perceived contribution to product or service
Internal social relations	Opportunity for social contact, recognition for achievements, provision of interlocking and mutually supportive roles, opportunity to lead or help others, team morale and spirit, small-group autonomy and self-regulation
External social relations	Job-related status in the community, few work restrictions on outside lifestyle, multiple options for engaging in work (eg flexitime, part-time, job share, subcontracting)
Career path	Learning and personal development, opportunities for advancement, multiple career path possibilities

Quality of working life criteria checklist (Davis & Wacker 1987)

Motivation is an important consideration in designing jobs and structuring the work organisation. An individual's effectiveness is influenced by their motivation; in a motivated state the individual is ready to perform better. Recognising the importance of motivation, most job designers draw heavily on theories such as that expounded by Maslow (1954). Maslow suggested that people have a hierarchy of needs and that five major needs direct their actions. These needs are:
- physical needs – consisting of primary needs such as food, water and sex;
- safety needs – including protection from physical harm, ill health and economic disaster;
- social needs – including the desire to feel part of a group and the need to establish a position in relation to others within the organisation;
- achievement needs – consisting of the need for self-respect and a feeling of competency; and
- self-actualisation needs – including the need to achieve one's fullest potential in terms of self-development and creativity.

Maslow's theory on the hierarchy of needs is a dynamic one. His theory states that at any given time any one need may be operating. However, an individual responsible for job design can apply the theory by concentrating on the physical and safety needs first – providing appropriate rest break schedules, suitable heating, lighting, ventilation and opportunities to eat and so on. If these lower grade needs are not met the designer cannot possibly hope to meet the higher grade needs.

Before designing the job itself, six separate decisions should be made:

- **What tasks will be completed by the workforce?** The organisation should first review workplace technology such as equipment, robots and materials and determine which technical and organisational tasks remain for the workforce to complete. Technical tasks contribute directly to the production system. An example of this would be an operator who is required to remove boxes of pre-packed food from a conveyor belt and pack them into an outer case. The production system will have cooked the food, placed it in the container, sealed it and passed it on to the operator for packing. Organisational tasks involve the planning, training, problem-solving and co-ordination which supports the production process.

- **How will individual tasks be structured?** Decisions have to be made with regard to the complexity of each individual task. The job designer has to decide if an operator will be given a series of complex sub-routines to complete, or a fragmented operation which they will repeat frequently.

- **How will a range of tasks be grouped together?** Consideration should be given to the grouping of tasks and whether it will be done simply to facilitate the process or to provide the operator with a meaningful and interesting range of opportunities. The decisions will be influenced by constraints such as location of tools, equipment and controls. Such constraints will limit the range of movement of an operator or group of operators from certain work areas and workstations and the variety of work they can undertake.

- **How will the tasks be assigned to individuals?** Consideration has to be given to the abilities and limitations of the individual and these will be influenced by their personal characteristics, training and experience. The organisation's in-house approach to work distribution and worker responsibility will also have to be taken into account.

- **How will the work be co-ordinated?** Co-ordination of work creates a link between all individuals within an organisation. The generation of this link can be achieved through open communication, formal reporting procedures, meetings, team identities, hierarchical management and supervision, and informal personal relationships.

- **How will individuals be rewarded for their work?** Rewards can be financial but they can also come in the form of promotion, opportunities for further training or additional responsibility. However, performance has to be measured to allow for the adequate reward for work. Performance can be evaluated through supervisory or managerial evaluation, tests, achievement of goals and targets, piecework, educational achievements and so on.

Having made these decisions, consideration should be given to the mechanics of job design. This chapter discusses the characteristics of jobs and work organisations which determine the individual's level of satisfaction as well as their physical well-being.

Repetition

As a consequence of the trend towards specialisation, certain organisations have designed tasks which are highly repetitive. It is understandable with large organisations, producing numerous products in a given timescale, that the most 'efficient' way of ensuring that the desired number of items is produced to the appropriate standard is to break tasks down into small sub-components and train an operator to complete one sub-component. By doing so the organisation can virtually guarantee that the operator will become highly skilled in the completion of that operation and will almost always produce high quality work. This was the approach taken by car manufacturer Henry Ford in the 1920s. Such simplification of jobs is not unique to production line operations. It is also apparent in office work where individuals are required to sit for long periods inputting information using a keyboard. The data-entry operator is a prime example of an office-based operator involved in highly repetitive work. As indicated in the chapter entitled 'Upper limb disorders', such rapid repetitive movements can have serious implications for an operator in terms of developing a ULD.

Putz-Anderson (1988) has provided a system for determining whether a task can be classified as 'high repetitive' or 'low repetitive'. He has suggested that a task with a cycle time of less than 30 seconds, or one where more than 50 per cent of the cycle time involves performing the same kind of fundamental cycle, can be classified as high repetitive. (A fundamental cycle is a work cycle that has a sequence of steps or elements that repeat themselves within the cycle.) A long cycle time does not necessarily mean that the job is not repetitive. For example, the cycle time for a garment packer to pack three pairs of briefs into a plastic pack may be 45 seconds which might suggest, using Putz-Anderson's criteria, that it would not be classified as high repetitive. However, as three pairs of briefs are packed, the fundamental cycle time is actually 15 seconds and the job should therefore be classified as high repetitive.

When designing any task, the ideal aim should be to offer as much variety as possible so that the work is not focused on one single part of the body, resulting in a possible overload of that structure. If there is greater job variety, the operator is also less likely to become bored or feel isolated. If the job content cannot be changed, the work could be reorganised. A programme of rotation could be designed which would allow operators to move away from repetitive tasks to other less repetitive or less demanding operations. The aim should be to provide the individual with an opportunity to recover from the stresses of the first operation. To achieve this, the subsequent activity should be qualitatively different to the first, so that a second set of muscle groups is used. This allows the muscles used during the previous operation to rest and recover.

Job rotation should not be undertaken without full consideration of all the tasks to be included in the programme and the activities involved when completing each task. It is only by doing so that each of the tasks can be included in a programme which offers regular recovery periods. If an operator is moved through a series of equally repetitive, short cycle tasks, they will not benefit because they will experience similar levels of musculoskeletal loading during each operation.

If operators are to be rotated, they should be fully trained in each of the tasks they are to complete during a rotation programme. Without the appropriate training, they may experience further problems as they attempt to complete an unfamiliar operation.

Worker flexibility – or 'multi-skilling' – is a refinement of job rotation. With this approach the worker is trained to perform a number of jobs and is then deployed in different areas as circumstances dictate. Worker flexibility normally generates an increase in remuneration. It is considered that this approach offers a more varied and interesting range of work experiences than would apply with traditional job rotation. Such multi-skilling is of particular benefit to the organisation during sickness or holiday absences as operators possess the skills which enable them to move from job to job and cover for absentees.

An alternative way of tackling the issue of repetition is to consider 'job enlargement'. This approach increases the number of individual tasks an operator has to complete without a change in responsibility. Other benefits of job enlargement include reduced storage area for work in progress and reduced handling time as operators do not need an allowance for packing and repacking items as work is moved from one area to another. However, operators should not be allowed to divide their newly enlarged tasks into a series of sub-tasks and complete each sub-task repetitively for a period of time before moving onto the next sub-task. This approach undermines the purpose of job enlargement.

'Job enrichment' differs from job enlargement in that it gives the operators more responsibility over their own work and possibly a greater say in the decision-making process, particularly in respect of their own team or work group or their work activities. Quality control, testing of products and re-work are examples of job enrichment. It can be introduced successfully in areas where there are (semi) autonomous working groups. The development of such groups increases co-operation between group members, encourages a more positive approach to work, and reduces boredom, monotony, isolation and absenteeism.

Work rate

Typically, the speed at which an operator is expected to work is dictated by the production capacity of the equipment or machinery they are using. Organisations expect the machinery to pay for itself over a period of time – and want it to work at maximum capacity. However, increasing the speed of the operation serves only to increase the stresses placed on the operator. The aim should be to set the speed of operation to fall within the operator's optimum – not maximum – level of functioning. This can be compared to the physical states experienced when running or walking. Most people can run for a short time but will find it demanding and tiring. Walking at a comfortable pace, however, will probably allow a greater distance to be covered. It is the same in the workplace. Operators can probably work faster but to do so increases their fatiguing rate and the likelihood that they will suffer strain or injury.

If an acceptable work rate is agreed, it is important that it remains consistent throughout the course of the shift and the working week. Operators should not be expected – or allowed – to work faster at certain points in the day or week. Working faster may be a result of a change in demand, or the operator choosing to work that way. An operator with targets to meet by the end of the day may elect to work faster in the morning, meeting up to three quarters of their target so that they can have a slower, more relaxed afternoon. Bursts of activity such as this result in residual fatigue which standard rest breaks cannot counteract. Standard, regular breaks are designed to allow a rest from activities which are performed at a consistent pace across the course of a shift. Peaks and troughs in activity should be controlled as far as possible.

Workload

The workload is the amount of work an individual is expected to complete within a given timescale. Whatever level this is set at, there is usually one crucial factor which should be recognised: all operators require a period of time to become 'work hardened' or 'task fit'. This is similar in nature to the level of fitness required of a runner before entering a marathon – an athlete would not expect their body to perform at its best without a gradual build-up, or acclimatisation, period. A similar approach should be taken at work, particularly with new employees or those returning from sickness or holiday leave. Even long term employees will have lost some of their task fitness after a two or three week holiday. Therefore, to avoid overload of the musculoskeletal system, they should be allowed a gradual build-up period to become accustomed to the demands of the work again. New employees and those returning to work following an absence are particularly vulnerable to the development of work-related injuries.

This level of 'vulnerability' is also apparent in workplaces where work demands change suddenly as a result of seasonal adjustments or increased customer demand. It is not unknown for an organisation to double or treble its workforce's workload overnight because of an increase in demand. Unless the organisation takes on temporary workers to assist in meeting the increased demand, or the workload is increased gradually, it will probably be faced with a sudden increase in complaints of ULD-type symptoms from the permanent workforce.

Increasing the workload to the point of overload (ie where demand outstrips capability) may lead to mistakes, errors or accidents as the operator develops a 'trade-off' strategy where the work is completed but with compromises. An example of this would be an assembly operation where an operator is supposed to fit four nuts to a part but because of the overall workload they elect to fit three to keep up with the speed of production. Attempts by the operator to rectify any mistakes or errors will themselves increase the workload further as they have to maintain production levels while at the same time trying to resolve problems relating to previous tasks.

Rest breaks

Rest breaks are usually a major source of discussion within organisations. Generally, the focus is on the amount of time operators are allotted during a shift. However, frequency rather than duration of breaks is important. Short, frequent breaks are more beneficial than longer, irregular breaks. The aim should be to allow the person to stop work before they actually start to fatigue – so that when they stop work they will *rest*. When work is resumed they can pick up where they left off. If, however, operators work for long periods of time without a break they will start to fatigue and their performance will deteriorate. When they finally stop work they will be *recovering*, or *recuperating*, from the stresses of the activity and not resting. When they resume activities, they have to build back up to their original performance level.

It has been suggested (Grandjean 1988) that as a minimum in moderately heavy jobs, operators should have a break of at least 10–15 minutes both morning and afternoon, in addition to their lunch break. However, a few minutes' rest every hour has been shown to reduce fatigue and improve concentration. This is particularly effective in repetitive assembly line-type jobs.

For VDU operators, it is recommended that a break of 5–15 minutes in every hour is desirable if the work requires high levels of concentration and is of an extended duration (Pheasant 1991). If the working conditions are appropriate and the task is interesting, five minutes in every hour would be sufficient. For boring or stressful VDU work, 15 minutes might be more appropriate. As a rough guide, an average of 10 minutes in every hour is advisable.

In addition to designing work so that the operator can stop altogether during standard rest breaks, consideration should be given to introducing 'micro-pauses'. These are very short rest periods, possibly lasting only a few seconds, which provide a brief respite from the stresses of the overall task. For example, a keyboard operator would stop to read their screen at regular intervals (perhaps every 10 minutes); or a welder feeding parts to welding machines would be allowed to stand and wait for the welding operation to finish before removing the part, instead of using the time to complete another operation.

Bonus and piece-rate systems

To ensure a high level of production, many organisations offer bonus or piece-rate payments. However, these simply encourage operators to work at speeds which move beyond their optimum level of functioning towards their maximum. Such systems increase the rate of repetition and sometimes decrease the regularity with which breaks are taken as operators work through them. Alternative systems of rewarding operators should be considered.

Overtime

Many operators view overtime positively as it provides an opportunity to increase their 'take home pay'. However, problems associated with overtime should not be overlooked. Overtime obviously increases an individual's exposure to their working environment. This extended exposure may cause problems if the normal shift involves repetition, force, deviation of the wrists, static muscle work, cold temperatures and so on. Overtime may also reduce the possible available recovery time. Therefore, overtime should be viewed not only in terms of the benefits for the organisation and the individual but also in terms of the disadvantages. If an individual is involved in a demanding, repetitive operation during a normal shift, it may not be appropriate to allow them to work overtime.

Work schedules

Schedules of work will influence an operator's attendance, motivation and commitment. The compressed working week results in longer working days on Monday to Thursday but permits the operator to have a three-day weekend. With flexitime the operator is required to be present at work during core times but outside these periods may choose their own work schedule. Such a system gives employees greater control over their working hours and thus increases motivation. Job sharing, where two share a single job, allows the individuals concerned to achieve goals related to both family and work life. The organisation also benefits from this arrangement because it can use the talents of two people, as well as having access to those who might not necessarily be motivated to go to work at all due to family or other commitments.

If using such work schedules, the organisation should ensure that operators are not overloaded due to a shorter working day, or are not working through breaks to make up time on the flexi schedule.

Shiftwork

Certain types of employment require job-holders to work when most others are at home asleep (eg nurses, security guards, bakers). In the manufacturing and service industries, the 24-hour period is usually divided into three eight-hour shifts (6.00 am–2.00 pm, 2.00 pm–10.00 pm and 10.00 pm–6.00 am). Most people dislike irregular, antisocial working hours. About two-thirds of shiftworkers suffer ill health of one form or another and about a quarter eventually give up shiftwork because of this ill health.

One of the biggest problems with shiftwork is sleep disturbance, mainly because the work disrupts natural body rhythms. It takes about two weeks for the night-worker's body rhythms to fall in step with the 'unnatural' day-sleep night-work pattern. However, many shifts change back to daytime work after two weeks. Once the operator makes the change to day shifts their normal body rhythms reassert themselves. Once back on night shift the process starts all over again. As a result, for much of the time the operator will be working when their body tells them they should be sleeping, and sleeping when their body tells them they should be active. The sleep disturbance is exacerbated by the fact that as the shiftworker tries to sleep during the day, the rest of the world is carrying on noisily around them.

As a result of lack of sleep, the fatigued operator can become irritable and lack their usual drive. Night-workers are more likely to suffer psychological problems such as depression, difficulty in concentrating and loss of self-esteem. There can be gastrointestinal symptoms such as loss of appetite, indigestion, constipation and, in the longer term, peptic ulcers. There may also be an association between shiftwork and coronary heart disease.

At present, evidence from studies suggests that significant improvements can be gained by adopting a slowly rotating shift system.

Other organisational issues

Automation

Automation is a fundamental part of progress in modern businesses. The movement towards it has been encouraged by the increasing cost of the human workforce, the existence of operations which are difficult or hazardous for the individual to complete, the lack of sufficient appropriately trained labour and the desire for increased competitiveness through reduced costs. However, factors such as the cost of automation, the lack of appropriate automation technology, internal organisational forces such as unions, and the lack of a market to necessitate the increase in production, may influence the development of automation. Few organisations have fully automated systems. Many use partial automation coupled with human input or effort. In these cases, it is important that the people and machines complement each other.

When the operator is expected to interact with equipment during a process, its design should not overly stress them physically or mentally. Ideally, the equipment designer should be able to develop a 'win-win' situation where the automated process eliminates the dangerous, repetitive or mundane tasks previously completed by the operator. The biggest drawback of automation is machine-pacing, which removes the control an operator has over their speed of operation and method of work. Some studies have shown that operators who prefer machine-paced work are less intelligent and more 'humble', practical and group-dependent. Those who prefer self-paced work are more intelligent, assertive, imaginative, shrewd and self-sufficient. Findings such as these underline the importance of basing recruitment and selection decisions on an individual's 'suitability' for the task.

Certain job characteristics should be considered prior to the introduction of new manufacturing technology. These will influence the performance of the operators and the new technology as well as the levels of job satisfaction and dissatisfaction experienced by the operators. Characteristics to be taken into account include:

- **Control** – how much control does the operator have over the process in terms of how a job is completed, and, what happens at the boundaries where the system takes over or where another operator assumes responsibility?
- **Cognitive demand** – most advanced manufacturing technology requires the attention of flexible, sophisticated operators who can interact with the system as they watch over it, diagnosing and solving problems as they occur. Production systems demand rapid responses to problems and the selected operators have to be capable of reacting in the appropriate manner.
- **Responsibility** – clear lines of responsibility for the system and any output should be identified in advance.
- **Social interaction** – what are the opportunities for, and what is the quality of, interaction between people associated with the introduction of the new manufacturing technology?

The biggest mistake made by organisations when introducing new technology is to ignore, or give insufficient attention to, implications for both the individual and the organisation as a whole. The success of the introduction is tied up with these elements. Organisations commonly assume that previous infrastructures will be suitable for the change in technology. This may not necessarily be the case.

Managing change

Change does not happen in a vacuum. It has an impact on everyone within the organisation and if not managed properly can be a source of employee stress. Change produces uncertainty, feelings of lack of control and, perhaps, an increased workload. It is essential that organisations construct a means for planning, implementing and monitoring change. Planning is required if the organisation is to determine exactly what should be changed, avoid unintended ripple effects, overcome resistance to change and harmonise changes between areas. The implementation process overlaps with the planning process but it should also identify:

- the change agent (ie someone with the expertise to bring about the change);
- the change model (ie the means by which the change will take place); and
- any constraints on implementation (including resistance, leadership climate, individual characteristics and the formal organisational design, eg hierarchical structure).

One of the main priorities should be to introduce the concept of change to the workforce in advance of the changes themselves. This can be accomplished through briefings and discussion groups, which give employees the opportunity to understand why change is necessary and what the consequences would be if the changes are not introduced, as well as the chance to ask questions about how the changes will affect them. If an employee's apprehensions are dealt with in the early stages, the climate for change will probably be much more favourable.

Many organisations face resistance to change from both individuals and groups. The resistance is an attitude that is determined by individual characteristics (eg perceptions and past experiences), organisational factors (eg lack of information and participation) and the content of the change (eg what might be lost by the change). The resistance results from an evaluation of losses and gains. Perceived negative outcomes typically include economic loss, uncertainty, inconvenience and disruption to social life.

Commonly used approaches for overcoming resistance to change include:

- **Knowledge and communication** – providing employees with information on the reason for and consequences of the change.
- **Participation and involvement** – employees have a greater sense of control if they participate in the change process. In addition, the organisation benefits from their first-hand knowledge of the activities.
- **Provision of incentives for compliance** – incentives can be financial, or non-economic such as greater flexibility in working hours or greater autonomy.
- **Provision of empathic support** – employees are given an opportunity to discuss their fears with a sympathetic listener and are offered tutoring and coaching opportunities which enable them to develop increased levels of competence and confidence.

Having introduced the changes, the organisation should strive to reinforce new skills, knowledge and attitudes through appropriate managerial and supervisory support.

No process of change is complete without an evaluation. Progress should be measured and followed up to assess the effectiveness of the changes. Feedback will enable effort and behaviour to be directed towards the desired outcomes. Since positive results may take some time to appear, evaluations should be carried out over a lengthy period. The evaluations should be viewed as an integral part of the change process. To achieve an overall balanced work system, adjustments or modifications will invariably be required and it will probably take some time before many changes are fully implemented.

IN SUMMARY

Job design
- Motivation is of primary importance in job design.

Repetition
- Repetitive tasks should be avoided.
- Where repetitive tasks exist, job rotation should be used to reduce the risk of injury, boredom and monotony.
- Each task included in a rotation programme should be qualitatively different from the preceding and succeeding operations.
- Multi-skilling is a refinement on job rotation where the operator is trained to complete any operation within a given area. This offers greater variety than traditional job rotation.
- Job enlargement can also be used as a means to combat the effects of repetitive tasks. This increases the number of individual tasks an operator completes without a change in responsibility.
- Job enrichment gives the operator more responsibility for their own work and a greater say in the decision-making process.

Work rate
- Operators should be encouraged to work at a consistent rate throughout the course of a day and from day-to-day. Peaks and troughs in activity should be avoided.

Workload
- Operators should be given an opportunity to become accustomed to the demands of any new or recently altered task before they are expected to produce a set level of output. This enables them to develop a level of 'work hardening' or 'task fitness' which will reduce the risk of injury.

Rest breaks
- Rest breaks should be made available to all operators at regular intervals.
- Short, frequent breaks are more effective than longer, irregular breaks.
- Micro-pauses, lasting only a few seconds, should be designed into tasks if possible.

Bonus and piece-rate systems
- Bonus and piece-rate systems of payment are not ideal. They encourage operators to work faster, for longer periods of time and with few, if any, rest breaks.

Overtime
- Overtime extends an individual's exposure time to their working environment. If this environment is a stressful or physically demanding one, overtime will increase the likelihood that problems will occur.

Work schedules

- Schedules of work influence attendance, motivation and commitment.

Shiftwork

- It takes about two weeks for a night-worker to become used to the unnatural day-sleep night-work pattern.
- Night-workers are more likely to suffer from gastrointestinal symptoms and psychological problems than day- or evening-workers.

Automation

- The effects of automation on the workforce should be considered (eg changes in levels of control and responsibility, changed demands and interactions with other operators).

Managing change

- Any changes should be managed and introduced with care – this process will influence the level of acceptance among the workforce.
- Once changes have been made, they should be evaluated to ensure that they have become part of a balanced work system.

REFERENCES AND FURTHER READING

Bridger R S, 1995, *Introduction to ergonomics*, McGraw-Hill, Singapore.

Davis L E and Wacker G J, 1987, Job design, in Salvendi G (ed), *Handbook of human factors*, John Wiley, New York.

Grandjean E, 1988, *Fitting the task to the man* (4th edition), Taylor & Francis, London.

Haslegrave C M, Wilson J R, Corlett E N and Manenica I, 1990, *Work design in practice*, Taylor & Francis, London.

Kroemer K and Grandjean E, 1997, *Fitting the task to the human* (5th edition), Taylor & Francis, London.

Lund R T, Bishop A B, Newman A E and Salzman H, 1993, *Designed to work: production systems and people*, Prentice Hall, New Jersey.

Maslow A H, 1954, *Motivation and personality*, Harper & Row, New York.

McKenna E, 1994, *Business psychology and organisational behaviour*, LEA, Hove, UK.

Monk T H and Folkard S, 1992, *Making shiftwork tolerable*, Taylor & Francis, London.

Moon S D and Sauter S L (eds), 1996, *Beyond biomechanics: psychosocial aspects of musculoskeletal disorders in office work*, Taylor & Francis, London.

Pheasant S, 1991, *Ergonomics, work and health*, Macmillan Press, London.

Putz-Anderson V, 1988, *Cumulative trauma disorders: a manual for musculoskeletal diseases of the upper limbs*, Taylor & Francis, London.

Statt D A, 1994, *Psychology and the world of work*, Macmillan Press, London.

Wilson J R and Corlett E N (eds), 1995, *Evaluation of human work*, Taylor & Francis, London.

MANUAL HANDLING

Introduction

Generally speaking, when people think about 'manual handling' they picture a heavy load being lifted by a stocky man in an industrial setting. This naive view of manual handling operations is one reason behind the serious problem of manual handling injuries at work.

In order to control the number of handling accidents and injuries it should be recognised that:
- women as well as men may be involved in the handling of loads;
- it is not only the weight of a load which is problematic; and
- manual handling is not restricted to industrial sectors – people are involved in handling activities in a variety of environments, including offices, hospitals and schools.

There is a tendency to overlook much of the manual handling carried out 'behind the scenes' by individuals such as post room workers dealing with a variety of incoming and outgoing mail; canteen workers moving cans, pots and bowls full of liquids and foods during meal preparation; and porters and maintenance personnel who are expected to move an unpredictable variety of objects in the course of a working day. Frequently, the manual handling elements of work are unrecognised and those carrying out the work are untrained and unaided.

Figures provided by the HSE (1998) show that 36.5 per cent of the accidents reported to enforcing authorities every year are due to the manual handling of loads (see below). Although many of these 'accidents' may appear to have a sudden onset following the movement of a single object, they are more likely to be cumulative injuries resulting from months or years of over-use or abuse of the body.

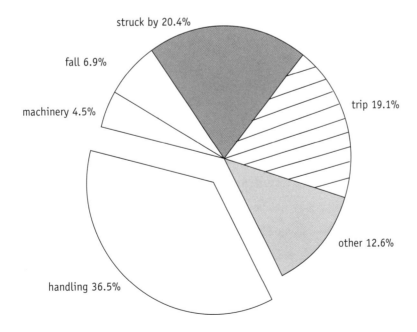

Types of accident reported to enforcing authorities

Reproduced from 'Manual Handling Operations Regulations 1992: guidance on Regulations' (HSE 1998)

The HSE statistics indicate that back injuries represent 49.3 per cent of all injuries received following a manual handling accident (see below). The outcome of all the injuries can be physical impairment, with the possibility of permanent disability.

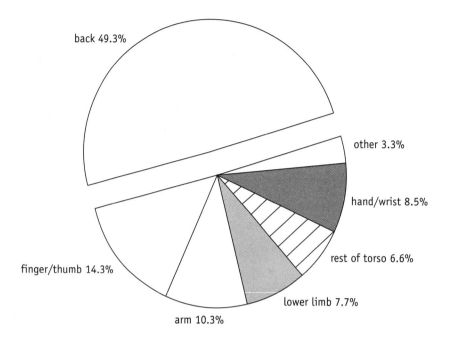

Sites of injury following manual handling accidents

Reproduced from 'Manual Handling Operations Regulations 1992: guidance on Regulations' (HSE 1998)

The cost of work-related accidents and ill health

In addition to the figures reflecting the number of accidents resulting from manual handling operations and the percentage which affect the back, the 'cost' factor can be easily illustrated. Employees need to be made aware that a handling accident can have an impact on quality of life as well as having financial implications.

The affected person may not enjoy life as before – and could find that they have changed from being a healthy and happy individual to being incapable of putting on their own shoes.

From the financial point of view, there is a loss of income in the short term as the individual recovers from the injury. In the longer term they may not be able to return to the same job – or to work at all. There may be additional expenses such as drugs and other treatment, and alterations to their home or car to ease their discomfort or accommodate a disability.

The costs to the employer are more clearly defined. The most immediate outcome following an accident in the workplace is loss of output. This can result from the absence of the injured party or others stopping work to assist their injured colleague. Equipment could have been damaged during the accident which may also affect production. Falling behind schedule can be remedied by overtime but this in itself is costly.

The organisation may be unable to meet its contractual obligations as a result of falling behind schedule (and financial penalties are not uncommon). In addition, in a competitive marketplace, the loss of the client's goodwill or damage to the organisation's reputation should not be overlooked.

Assigning people to investigate the accident and prepare a report, along with organising the repair or cleaning of the accident area, also have cost implications.

The costs directly related to the injured party will include additional administration and the recruitment and training of replacement workers. The employer may be responsible for providing extra medical treatment such as physiotherapy and a level of compensation may be required. Although the employer's liability insurance will cover such compensation payments and the associated legal fees, these extra costs will be reflected in increases in the annual premiums. Finally, there could be fines or other penalties (including imprisonment) imposed, again with associated legal costs.

It has been estimated that the indirect and uninsurable costs of accidents (which include investigation costs, recruitment and training and loss of goodwill) are at least eight times the direct insured costs (which include sick pay, repairs and compensation payments).

Tackling the problem of manual handling

It is unlikely that the problems associated with manual handling operations will ever be completely eliminated. For as long as people use their own body to lift, carry, push, pull, throw, drop or support an object there will always be the possibility that they may injure themselves. No handling operation can, or should, be viewed as 100 per cent 'safe'. However, the risks associated with a manual handling operation can be reduced so as to minimise the possibility of injury. Following sensible guidelines should help to reduce risks – such guidelines are presented throughout this chapter.

Positive steps to tackle the problems associated with manual handling include:
- identifying areas where any objects are manually handled;
- acknowledging that there may be the potential to cause harm; and
- recognising that it is within both the employer's and employee's power to combat potentially damaging manual handling problems.

The Manual Handling Operations Regulations 1992 provide employers with a means to assess and control the risks associated with manual handling activities.

Manual Handling Operations Regulations 1992

The Regulations came into force on 1 January 1993. They implement European Directive 90/269/EEC on the manual handling of loads.

The Regulations provide a clear three-tier hierarchy of measures which present the employer with a step-by-step guide on how to tackle the risk-bearing operations. The measures are:
- any hazardous manual handling operations should be avoided – so far as is reasonably practicable;

- a suitable and sufficient assessment should be made of any hazardous operations which cannot be avoided; and
- the risk of injury from these operations should be reduced – so far as is reasonably practicable.

The employer's duty to avoid or reduce the risk of injury will be satisfied within the confines of 'reasonably practicable' if they can show that the cost of any further preventive steps would be grossly disproportionate to the further benefit that would accrue from their implementation.

Interpretation

The Regulations provide a specific interpretation of the terms 'injury', 'load' and 'manual handling operations':

- **Injury** refers to injuries sustained by any part of the body – not just the back. Examples of the types of injury which can occur include those caused by picking up a load which is too heavy, by cutting the hand on a sharp edge, or by dropping the load onto the foot as a result of slippery packaging.
- **Load** refers to a 'discrete moveable object' such as a box, printer, product component, animal, person or material supported by a fork or shovel. It is important to recognise that an implement, tool or machine is not considered to be a load when in use for its intended purpose. An example of this would be a large welding gun used in a car body factory. Although suspended overhead these guns are renowned for being extremely heavy and difficult to manoeuvre. In this situation, the use of the welding guns would not be assessed under these Regulations. However, should a consignment of such guns arrive at the plant on a lorry and have to be unloaded by people, they would then be viewed as a load.
- **Manual handling operations** refer to the movement of a load by human effort (as opposed to using a crane or forklift truck). This effort can be applied directly, such as when gripping a load with the hands, or indirectly, as when tying a rope around a load and pulling on it. The hands do not have to be used at all during the handling of the load – the operator can use other parts of their body, such as the foot or shoulder, to move an object.

The term 'manual handling operation' includes what would be viewed as 'typical', ie picking an object up from one place and locating it elsewhere. It also includes other activities such as:

- supporting the load without moving it;
- intentionally throwing or dropping the load; and
- pushing and pulling (which means the load does not have to clear the supporting surface).

Even if operators are provided with mechanical assistance to move the load, they will still be required to move, steady or position the load during its relocation. Therefore, the operation will still involve an element of manual handling.

Duties of employer – Regulation 4

Generally, the employer is expected to reduce all risk of injury so far as is reasonably practicable. Having introduced measures to reduce the risks, the employer is expected to monitor the interventions to see whether they have achieved the desired effect. If not, they must review their decisions and develop alternative strategies for risk reduction.

All assessments should be kept up-to-date bearing in mind that the balance within the reasonably practicable equation may alter as a result of changes in technology, reductions in the cost of handling aids or alterations to the workplace. Assessments should be reviewed if there is a change in the handling operations which could alter previous conclusions or if a reportable injury occurs.

Employers are obviously responsible for employees who work at their own site. But they are also responsible for employees who are working away from their own premises. Although it is difficult to exert control over the conditions at these other working environments, the employer does have control over the training given to employees and, possibly, control over the task and load. For example, if employees are sent to another site to carry out maintenance work, the employer can ensure that they pack their tools and equipment into several manageable loads rather than one or two very large loads. The employer can also ensure that employees are aware of the need to work at a suitable pace with regular rest breaks. Employers in control of premises at which visiting employees have to work have duties towards them under the Health and Safety at Work etc Act 1974, the Management of Health and Safety at Work Regulations 1999, and the Workplace (Health, Safety and Welfare) Regulations 1992, to ensure that the premises and plant provided are in a safe condition.

Anyone within the organisation can undertake the assessments on behalf of the employer. Whoever is given the responsibility of making the assessments should have a clear understanding of the types of handling operations performed, the range of loads which will be handled and the environments in which the handling operations occur. Outside assistance can be sought, particularly with problem areas, however, the final responsibility for the assessment lies with the employer.

It is recommended that employers do not ignore the valuable input of employees and their safety representatives. Those directly involved in handling operations are more likely to know exactly where the problem areas are – and may be able to provide recommendations for change.

Information from records kept by the employer can offer an insight into areas which may require immediate or more thorough assessment. Records from the medical department may show areas with a high incidence of injuries and personnel records may indicate areas with frequent absences. Further sources of information include those relating to productivity, quality, rejects and damage. This information should only be used to augment other assessment methods.

Assessments will be judged 'suitable and sufficient' if they address in a considered way all the handling operations carried out at the site. Employers should bear in mind that they can carry out 'generic' assessments which will pull together the common threads of many similar operations. A sample checklist is provided within the guidance to the Regulations to assist in the assessment procedure. The assessments should provide a profile of the risks to which employees are exposed and assist the employer in constructing preventive steps where necessary.

In general, the significant findings of the assessment should be recorded and kept for as long as they are relevant. However, the assessment does not need to be recorded if the task is simple and obvious and can be easily repeated or explained at any time, or if the handling operation is of low risk, straightforward and lasting only a very short time, and as such the time taken to record the findings would be disproportionate.

When a more detailed assessment is necessary, it is recommended that the assessor should examine the operation under five separate categories:

- **The task** – what the person is doing during the handling operation (eg lifting the load at a distance from the body, whether they are stooping, whether they are sitting).
- **The load** – the characteristics of the load itself, such as its weight, shape and packaging.
- **The working environment** – the area in which the person is handling the load (elements such as the floor surface, steps, temperature and space constraints should be taken into account).
- **Individual capability** – the personal characteristics of the person (eg height, strength, gender, age).
- **Other factors** – for example, the effects of using personal protective equipment (PPE) when carrying out a manual handling operation.

Having assessed the operations, the employer is expected to reduce the risks by adopting a similar structured approach in considering the task, load, working environment and individual capability. Methods for the reduction of risks associated with manual handling operations are discussed later in this chapter.

Finally, to assist employers in reducing the risk of injury, where the originators of loads (ie manufacturers or packers) are aware that the products are likely to be manually handled they may have relevant duties under the Health and Safety at Work etc Act 1974 and as such should give consideration to making a load easier to grasp and handle and to marking it with the weight and centre of gravity.

Duty of employees – Regulation 5

Employees are required to make use of any equipment provided for them, such as a handling aid, and use it in the manner instructed. Employees are also required to follow the appropriate systems of work laid down by their employer to make handling operations safer.

Exemption certificates – Regulation 6

In the interests of national security, the Secretary of State for Defence may exempt home forces.

Extension outside Great Britain – Regulation 7

The Regulations apply to offshore activities such as those connected with oil and gas installations.

Repeals and revocations – Regulation 8

These Regulations replace a number of outdated provisions which concentrated on the weight of the load alone. The provisions referred to are listed in Schedule 2 of the Regulations.

Manual handling injuries

It is important to remember that although the main concerns tend to focus on injuries involving the back, other less serious injuries resulting from crushing or trapping the fingers or feet or from tripping or slipping should also be addressed.

In addition, it is worth bearing in mind that it is possible to develop a ULD as a result of a handling operation (ULDs are discussed in detail in the chapter entitled 'Upper limb disorders').

The back

It is estimated that approximately 70 per cent of people will experience backache at least once in their life and this will last for a short period. However, it has also been recognised that the recurrence rate is very high – possibly 60 per cent of affected people will suffer from backache again within one year.

The primary reasons for the condition of the back deteriorating are: the natural ageing process; disease; illness; over-use; and general wear and tear. The pain which can be experienced takes many forms and can be acute, dull, located in one particular area of the back or all over it. Back problems can also create painful sensations from the buttocks down to the feet.

To appreciate why the back is so vulnerable to manual handling injury, it is necessary to have a basic understanding of how it is made up and how it functions (see below).

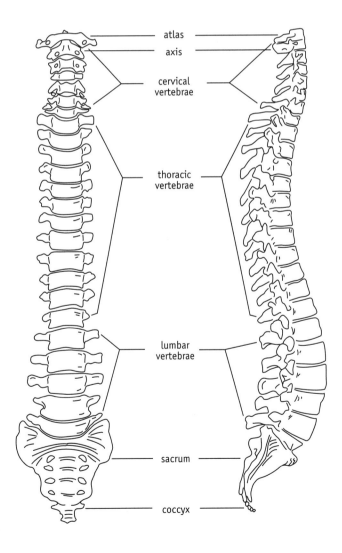

The spinal column

Reproduced from 'Hollinshead's Functional anatomy of the limbs and back' (Jenkins 1998)

In brief, the back consists mainly of:

- **Spinal cord** – this is a thick cord of nerve tissue which is enclosed by the spine. Together with the brain it forms the central nervous system.
- **Vertebrae** – these are the bones which act as the building blocks of the spine and they can be damaged by impact injuries such as those resulting from a heavy object dropping on the back from a height.
- **Intervertebral discs** – these are the 'shock absorbers' positioned between the vertebrae and they give the spine flexibility (see below). The discs are made up of a fibrous outer band with a fluid-filled centre. When a person stands or sits upright with an 'S'-shaped spine, downward pressure is exerted evenly over the surface of the discs. If a person bends over at the waist to pick up an object, the downward pressure will be exerted unevenly over the surface of the discs, with the loading being greater on the front edge of the disc than at the rear. At the same time, the viscous fluid inside the disc will be squeezed towards the rear of the disc.

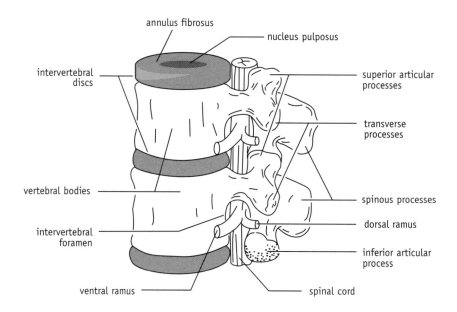

The discs

Reproduced from 'Hollinshead's Functional anatomy of the limbs and back' (Jenkins 1998)

Over time, the fibres on the rear section of the outer band may begin to tear and a bulge may form. The tear and bulge will become bigger if pressure is applied repeatedly and eventually the fluid from the centre of the disc may be pushed out, pressing on nerves which will cause the individual to experience pain. This is called a prolapsed (or slipped) disc.

Disc injuries are considered by some to be the most serious back injury and have been reported as being associated with 95 per cent of back injuries. However, it has been suggested that more realistically only five per cent of back injuries involve damage to the discs. Irrespective of the statistics, the adoption of more appropriate lifting techniques will help to avoid this problem.

- **Ligaments** – these are gristly straps which stretch between bones, holding them together. They mainly control the direction of motion, provide passive resistance and limit movement towards the end of the normal range.
- **Tendons** – these are the means by which the muscles are attached to bones.
- **Muscles** – these are found in pairs on either side of the spine. They provide the main stability of the vertebral column and provide passive resistance to spinal movement, the resistance increasing as they are stretched.

The ligaments, tendons and muscles are susceptible to injury as a result of twisting and stretching, particularly if carried out repeatedly over a long period of time. Injuries to these soft tissues are probably responsible for the majority of back injuries reported.

Hernias

Hernias are often associated with manual handling because the lifting action is accompanied by an increase in pressure within the abdominal cavity. The abdominal wall can have weak spots or gaps and sometimes during the lifting operation – particularly when the body is bent at the waist, compressing the internal organs – a loop of intestine may be forced out through a weak spot. A hernia is the protrusion of this loop through the abdominal wall. Poor lifting techniques which involve the operator bending at the waist are known to increase the risk of hernias.

Other injuries

Other injuries may include cuts, bruises and fractures. They typically result from poor housekeeping where tripping hazards are present, from poor lighting, from dropping or spilling of the load or from crushing or entrapment by the load.

Mode of movement

The manner in which a load is moved will determine how the body is stressed, how quickly it fatigues and the possible injuries which can occur. Lifting an object results in increased compressive force on the spine which can affect the discs, ligaments, tendons and muscles.

When pushing and pulling loads, both the abdominal and back muscles are called into play, thereby increasing the reactive compressive force on the spine. Pushing in particular increases the loading on the shoulders and breathing is made more difficult by the subsequent rigidity of the rib-cage.

Once an object is supported in a fixed position or carried across a distance, static muscle work is required. Such work arises from having to hold the muscles tensed for an uninterrupted period and is extremely fatiguing for the muscle groups concerned.

Avoiding the risks of a manual handling operation

Obviously, the surest way of reducing risks is to avoid the handling operation altogether.

Direct delivery

Some working environments lend themselves to the safe movement of loads by mechanical means. For example, loads such as flour, starch, grains and liquids could be pumped directly to the area of use, thereby eliminating the interim storage areas where loads are held temporarily before being rehandled and taken to the final area of use.

Other loads could remain on pallets at all times with the complete pallet being moved to the area of use via a forklift – rather than splitting the pallet and handling items individually.

Increase in volume

Many loads are on the borderline between 'tempting' and 'unmanageable' (eg 35 kg sacks of salt and 250 kg drums of detergent). Operators may attempt to move this type of load and as a consequence injure themselves. It may be worth considering the delivery of products in volumes which are impossible to move by hand. The only option open to operators in these situations would be to use forklift trucks. This is contrary to the view that making a load smaller makes it safer to move but it does remove the temptation experienced by some operators to 'have a go' and attempt to manage a load unaided.

Mechanical transportation

Conveyor lines, roller-beds and chutes will also eliminate the need for the operator to move an object manually from one area to the next.

Trolleys could be provided to eliminate the need to carry objects, however, they should be maintained to a satisfactory level. If the wheels or castors do not work properly because they have become gummed up with food or grease, have worn out, or are incapable of coping with the weight they are expected to move, it will be the operator who will have to overcome the resistance of the trolley by pushing harder or by using abrupt movements. Trolleys should be included in an ongoing maintenance plan. There should also be a quick response to operator reports that a trolley has stopped functioning properly. If a trolley requires the operator to push or pull with the hands above shoulder or below knuckle height, the movement of the load will be much more difficult and fatiguing. Ideally, the operator should be able to locate their hands around elbow height where the load can be managed more easily.

Handling aids

The provision of specifically designed aids should assist in handling operations. However, such aids can be expensive and serious consideration should be given to their design and use before they are transformed from paper to prototype.

When considering a handling aid the designer or purchaser should address issues including:
- **Will the handling device reduce the physical effort required to move the load?** It would be self-

defeating if the effort required to move the handling aid into position was similar to that required to move the load manually.

- **Will its introduction introduce any new problems?** If a handling device is being used alongside a production area where operators have to carry out other activities, it could become an obstruction, reducing their freedom of movement and interfering with their adoption of safe working postures.
- **Will it be easy to operate?** If a handling aid is too complicated or awkward to use, operators will tend to work without it when moving the load.
- **Will it function adequately in terms of speed and flexibility once introduced to the work area?** This is particularly important if the handling device will interact with a continually moving production line.
- **Will operators be satisfied with its use?** If operators are not satisfied with the design or introduction of a handling device, they may not use it.
- **Can operators use it comfortably when wearing PPE (particularly gloves)?** If the handling aid becomes difficult or more complex to work with because operators are wearing PPE they may work without it.
- **How much training will operators need?** This should be worked out before the handling aid is introduced so that the appropriate length of time is set aside. If appropriate training is not given, the safe operation – and continued use – of the handling device cannot be guaranteed.
- **Will the supplier provide the training?** If so, the employer should ensure that training is adequate.
- **What will happen to the device when it is not in use?** When it is not in use it may need to be moved out of the immediate area to provide additional work space. The device should allow for this movement.
- **Does the device include a facility to be locked off?** If the handling aid is free-moving (eg on an overhead gantry) efforts should be made to ensure that it cannot move around when not under the direct control of an operator.
- **Can the device be misused or abused?** A facility may need to be built into the handling device to prevent operators from abusing it.
- **Will the device damage or drop the object it should be moving?** For example, if a vacuum lifter is used in a dusty environment it might drop the load if the dust affects its contact with the load.

All these issues should be considered before designing or purchasing a handling aid. If they are not, it is possible that the (perhaps costly) handling aid will reside, unused, by the side of the work area. It is advisable to try out a device prior to full introduction into the workplace to determine whether there are any 'teething' problems.

Reduction in degree of handling

There are other methods of reducing risks involved with manual handling. In certain cases it may be possible to reduce the degree of handling involved. For example, many machines which prepare food products deposit the food into a container at the out-feed end. The operator commonly lifts the container and carries it to the weighing scales before carrying it to a pallet. Simply incorporating the scales into the bench at the out-feed end of the machine would remove the need to carry the load to the scales before depositing it on the pallet (which should be located close to the bench).

Of course, eliminating or reducing the degree of handling involved is not always possible, or practical, therefore other ways of reducing the risks should be found.

IN SUMMARY

- Direct delivery of loads, such as pumping to the area of use, will eliminate the need to handle the loads manually.
- Increasing the size of loads should reduce handling risks by making it impossible for the operator to move loads manually.
- Mechanical transportation of loads on roller-beds and trolleys could be used as a means of reducing the stresses of the handling operation.
- If trolleys are used they should be maintained at a satisfactory level.
- Handling aids should only be introduced after careful consideration of their design and use.
- Reorganisation of the workplace may reduce the number of times individual loads are handled.

The problems of manual handling

Typically, it is assumed that the weight of the object is the most problematic aspect of the manual handling operation (hence the common assumption that the weight will be implicated if an injury occurs). However, in many cases the weight of the load is not the key issue. If appropriate preventive strategies are to be constructed to combat the risks, a clear understanding of all contributory factors is required. The simplest way to examine the risks is to use the categories referred to earlier – task, load, environment and individual.

The task

When considering the task, the focus should be on what the person is actually doing as they move the load – attention should not be focused on the load itself at this point.

Posture

Posture is of great importance. The operator should achieve the appropriate posture before they attempt to take hold of the load. If they do not locate their feet so that they have a stable base, or if they do not take hold of the load in the best place using a strong grasp, they cannot hope to control the load properly once it clears the supporting surface. Appropriate lifting postures put the operator in a position of strength where they can control the load more easily during its movement and where they are less likely to injure themselves. Correct foot placement is the key to safe lifting. Training in good lifting techniques is discussed later in this chapter.

Twisting, while lifting, significantly increases the stresses placed on the lower back and is known to be particularly hazardous. Continual twisting is likely to act cumulatively rather than immediately in terms of aggravating the back. This action is typically seen when an operator moves an object between two surfaces located close together, such as between the end of a conveyor belt and a pallet. Instead of repositioning the feet to move closer to each surface in turn, the operator will often pick from the conveyor and twist at the waist to deposit on the pallet.

Untrained operators often bend over at the waist when picking up a load, or lean forwards when reaching to grasp a load at a distance from the body. Both actions increase the stresses on the lower back because once the operator throws their trunk forwards it becomes an 'unintended load' which adds its weight to the weight of the 'intended' object the operator is trying to move. In effect, two weights are being moved – the upper body and the intended load. This will be tiring, particularly if carried out repeatedly or for long periods of time.

Reaching

An operator may lift an object located at a distance from their body by stretching their arms out rather than moving the object closer. Lifting a load in this manner, when it is at arms' length, is five times more stressful for the back than lifting the same load close to the body. In addition, once the load is at a distance from the operator's body it is much more difficult to control and the friction between the load and the operator's clothing, which helps to hold the load in position, is lost.

The aim should be to have all the objects that need to be moved within a zone that can be reached easily by the sweep of the forearm when there is a 90° angle at the elbow. Of course, this is not always possible, therefore when working at a worksurface such as a desk, the heaviest and most frequently used items, such as reference folders, should be sited closer to the operator while the lighter and easier to move objects are stored on the outer edges.

Obstacles should be removed so that the operator does not have to reach over them to access the load. It is not uncommon to see operators reaching over boxes or pallets stored at floor level (which may be irrelevant to the operator and their task) when retrieving an object from a storage space directly behind them. In this situation, the risk of the handling operation is significantly increased. By moving close to the load the stress on the operator's lower back will be reduced.

Operators will often be seen reaching across a pallet, stillage or worksurface to retrieve or deposit a load rather than moving to a more appropriate position. Operators should be encouraged to move closer to the appropriate side of the storage container or worksurface so they are closer to the load. This will, of course, only work if there is enough space around the sides of the storage container or worksurface to allow them to move freely.

In some instances, simply tilting a storage container will bring a load closer to the operator, thus reducing the need for them to bend at the waist and reach forwards. If one side of a storage container can be removed it would make the retrieval of items stored towards the rear much easier. The operator responsible for the delivery of storage containers to production areas should ensure that they are delivered the right way round (there is, of course, no point in having a container with removable sides if the appropriate side is the furthest from the operator).

Working height

If the operator has to reach upwards to retrieve an object, stresses on both the arms and the back are increased. If the object is above head height, such as when reaching to an upper shelf or rack, operators cannot be sure that they have in fact grasped the uppermost object. Once they pull the load towards them they may find that something else, previously unseen, is situated above it. This other object may fall and injure the operator.

If items are to be stored on shelves or racks, they should be positioned in a manner which reflects their weight and the ease with which they can be handled. The greatest strength is found between knuckle and shoulder height and the heaviest or more difficult to handle items should therefore be stored on shelves or racks that fall within this range, preferably at waist height. Lighter or easier to handle objects could be stored on shelves which fall outside this range.

To ensure that items are consistently sited on the appropriate shelves it may be advantageous to label the shelves to indicate the range of items which should be stored at a particular level. These labels could also provide information on the weight of the object so that operators will know whether they need a second operator or a handling aid to assist in moving it.

For shelves which are located above shoulder height, the provision of an inexpensive platform or duckboard will make removal of objects easier – particularly for shorter operators. Should platforms be introduced, they should not be so small that they can be fallen from easily. Care should also be taken to prevent the platform from becoming a tripping hazard when other operators move around the area.

Stacking heights of goods on pallets should be controlled so that operators do not work above shoulder height. Pallets are typically stacked to a level well above head height which increases the risks involved in handling loads.

It is preferable for operators not to have to lift from floor level as this encourages stooping. Delivering loads directly onto tables or benches rather than onto the floor should be considered. Placing loads in raised storage containers or on benches with feet will allow the operator to position their feet under the supporting surface, which moves them closer to the load.

Locating a loaded pallet on top of another base pallet will raise it and make it easier for the operator to lift any loads on the pallet safely.

If it is not possible to raise the load off floor level, it would be preferable for the operators not to have to lift the load beyond waist height. If the load must go beyond this point it would be more appropriate to provide an interim surface on which to rest the load during the lift. This would allow the operators to change posture and grip so that they are in a stronger position when raising the load to its final position. Although it would appear that the amount of handling has been increased, it is better to make two safe lifts than one unsafe lift.

Walking distance

The longer an operator is in contact with a load the more likely it is that the handling operation will be fatiguing. If, having picked up a load, the operator has to walk a significant distance (> 10 m), the time involved in the handling operation is extended, and hence the stresses of the operation are increased. It is not uncommon for operators to be required to walk unnecessarily long distances due to poor workstation layout – objects are typically stored away from their area of use and their location has not been determined by frequency of use or overall weight. Simple improvements in workstation layout, such as locating the heaviest or most frequently used items closest to their area of use and placing lighter, infrequently used loads at more distant points, should reduce the risk of injury.

Pushing and pulling

The ease with which the operator can push or pull a load will be influenced by the condition of the floor, the type of footwear used and the nature of the load.

HSE guidelines for the pushing and pulling of loads suggest that a force of about 25 kg (approximately 250 Newtons) for men and about 16 kg (approximately 160 Newtons) for women for starting and stopping the load is acceptable, not forgetting that once the load is moving the force requirements change. The HSE suggests that a force of about 10 kg (100 Newtons) for men and about 7 kg (70 Newtons) for women for keeping the load moving is acceptable. These figures are based on the assumption that the hands are kept between knuckle and shoulder height. Pushing and pulling outside these zones may necessitate a reduction of the guideline forces.

For very large loads (or objects such as a sliding door) the operator could put their back against the load and use the leg muscles to exert the required force. Of course, by walking backwards, some vision impairment will occur.

Sudden movement

Should the load resist movement by the operator and then suddenly move, the operator may be unprepared for this and lose control of the load. The unpredictable movement of a load (eg a trolley wheel sticking in a pothole or product components tightly packed in a storage container suddenly becoming free) may put the operator at risk.

Seated lifting

Not all operators stand upright to handle a load. The seated operator carrying out handling operations faces additional problems. The operator will rely on the arms to do the work and may ultimately lean forwards to come within range of the object. Once they lean forwards there is the possibility that their seat may move backwards away from them. The seated lifter should not be expected to lift as much as someone who is standing. The HSE has recognised that special consideration should be given to the seated lifter. It provides a guideline figure of lifting no more than 5 kg for men and 3 kg for women when the person is sitting upright with their elbows at a 90° angle. In addition, the movement of loads from floor level should be avoided as the person will bend at the waist to reach the object. This will commonly be accompanied by twisting if a swivel chair is not in use.

Team lifting

Assigning two or more people to the handling activity may not eliminate problems. In fact, once other people join in the lifting operation additional problems are encountered. For example, having to ascend steps or slopes becomes a more complicated operation when several people are handling a load, particularly as each additional person may impede the view of their opposite number on the other side of the load. Further problems are encountered when the load's packaging is designed for one operator and only one set of handholds is incorporated.

When working as a team, the operators should ensure that the team members are of equal build and ability so that all are lifting under equal conditions. One of the team members should take control, co-ordinating the timing of the lift and the direction of movement. There should be sufficient room for all team members to move easily as they relocate the load. If handholds are available there should be enough for all team members. The total weight which can be safely picked up by a team does not increase in direct proportion to the number of people in the team. The HSE suggests that as an approximate guide the capability of a two-man team is two thirds the sum of their individual capabilities, and the capability of a three-man team is half the sum of their individual capabilities.

Work organisation

Obviously the speed at which an operator has to work, the length of time they have to work before having a break, and the degree of repetition, will influence if and when an operator will experience fatigue or discomfort. Even if the handling task cannot be changed in any way, the risk associated with the movement of a load can be reduced by the appropriate use of rest and rotation programmes. By effective use of these programmes, the operator's exposure time to the situation can be limited and so the risk of injury is reduced. A rotation programme should ensure that operators move between qualitatively different tasks (eg moving from loading boxes onto pallets at the end of a quiche production line to loading boxes onto a pallet at the end of a pie production line will probably not reduce the fatiguing effects nor reduce the risks of the handling operation).

The load

Weight

Often when attention is given to the risk presented by a load the focus is on its weight. However, this is only part of the picture. Making the load smaller and lighter might be one solution for reducing the risk. If the load is smaller it will be easier to grasp and handle. But consideration should be given to whether reducing the weight or size of the load will lead to an increase in the number of times the smaller loads have to be moved. In addition, handling smaller loads may result in operators bending down to lower levels such as when loading them onto a pallet.

If the weight cannot be reduced it may be beneficial to consider supporting the load in some way, particularly in a situation where a container such as a large jug is being used to fill other containers.

Shape

A bulky load will be more difficult to take hold of securely and safely. As a result, operators will have to use greater degrees of effort. Using more effort will result in them becoming tired at a faster rate. In addition, if the load is difficult to hold it is more likely to be dropped.

Size

If a load is large it may not clear the floor when lifted, which may result in it having to be dragged. If the load is lifted so that it does not have to be dragged it may interfere with vision. In both cases the risk of tripping or falling is increased.

Centre of gravity

The centre of gravity of an object is particularly important, yet it is a commonly overlooked factor. If an operator attempts to pick up a package which looks symmetrical, such as a cardboard box, and has not been advised that it contains an object such as a VDU screen, they may find that they are thrown off balance because the heaviest side is further away from their body than was anticipated. Even when a load is manageable it should be carried with the heaviest side towards the body to reduce the forces acting on the spine.

Sudden movements

If the load moves suddenly because it lacks rigidity (eg in the case of an unconscious person or animal) or because it has not been packed tightly within its outer packaging, operators will be at increased risk because they will be unprepared for the movement and the additional stresses it produces.

Grasping and moving the load

The outer condition of the load may make it difficult to move safely. If the edges are sharp or the contents are very hot or cold, operators may use gloves to protect their hands from cuts and burns. The use of gloves can impair dexterity making it more difficult to grip the object securely.

Making the object easier to grasp should be a major consideration. If the load itself cannot be altered it could be placed in a container with appropriate handholds – bearing in mind that adding a container increases the overall weight. Some seemingly 'lightweight' plastic containers (eg used to carry foods, nuts and bolts or documents) can add 2–3 kg to the total weight being moved.

The location and design of handholds will influence how easy an object is to move. The size or the centre of gravity of a load will influence where the handles should be located. The handholds should be designed so that they accommodate the operator's hands (which may or may not be protected by gloves). If the object is to be carried from the base it may be advantageous to cut finger wells into the bottom surface to prevent the fingers being trapped when the load is put down.

Packing material used within containers should not add unnecessarily to the weight of the load and should prevent the contents from moving around freely during their relocation. (Popcorn can offer a more environmentally-friendly alternative to polystyrene shapes or similar packing materials.)

If, as in the case of moving powders or liquids, packing cannot be used to prevent the shifting of contents, the containers should be filled as far as possible to limit the degree of movement. Should the contents still move around uncontrollably during relocation it may be safer to use a handling aid.

Labelling

If organisations are responsible for the packing of their own loads into containers, and these will be handled by other operators in the facility, a labelling system to identify the important characteristics of the load may be appropriate. Labels could carry information relating to weight and distribution of weight along with advice on how the load should be moved (eg 'with assistance', 'with a handling aid', 'with caution'). Colour coding and symbols could be used in place of a written description to provide a more immediate message that should be understood unambiguously by a greater number of people.

It may be possible to print appropriate information onto documentation used in the production process. For example, in a warehouse setting it would be relatively simple to incorporate information into a computer-generated picking list which would advise the operator prior to an item's selection that it may be heavy or bulky. The use of symbols would be particularly useful in this situation.

HSE guidance

The HSE has generated guidance for the lifting and lowering of loads (see below) which provides organisations with an indication of how acceptable their own working practices are in terms of the objects they expect operators to move during normal processing or production.

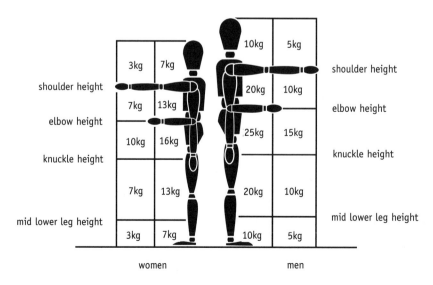

Guidance for load handling

Reproduced from 'Manual Handling Operations Regulations 1992: guidance on Regulations' (HSE 1998)

If this information is used as a basic guide for determining the acceptability of working practices, it should be borne in mind that these figures are based on the assumption that the operator is using both hands to pick up the load and that they are in a reasonable environment and are using appropriate lifting techniques. Reductions should be made to the guideline figures in certain situations, such as when the operator is twisting or carrying out a repetitive operation.

The guideline figures are based on infrequent operations of up to 30 lifts an hour with provisos that the task is not a paced operation, it provides opportunities for rest and recovery, and that the load is not supported for extended periods of time. It is suggested that the figures should be reduced by 30 per cent if the operation is repeated once or twice per minute, by 50 per cent if the operation is repeated five to eight times per minute and by 80 per cent where the operation is repeated more than 12 times per minute.

It is recommended by the HSE that the figures should be reduced by approximately 10 per cent if the operator is twisting by 45° and by 20 per cent if the operator is twisting by 90°. The HSE figures are based on an assumption that when twisting the feet remain stationary and the upper body (particularly the shoulders) is rotated to one side.

The HSE emphasises that these figures are only guidelines and are not limits. The figures can be exceeded but only if more in-depth assessment has shown that it is appropriate to do so.

National Institute for Occupational Safety and Health lifting guide

The US National Institute for Occupational Safety and Health suggests (1981) that the handling of loads should be viewed within the context of two limits: the action limit (AL) and the maximum permissible limit (MPL). The AL is a suggested level below which regulations and guidance are not necessary as the risk of injury is minimal. The MPL represents a maximum above which manual handling should not occur as the load is considered to be hazardous. The limits have been devised with reference to epidemiology, biomechanics, physiology and psychophysiology with the aim of reducing musculoskeletal injury rates, biomechanical compression forces and metabolic rates, and so that the loads do not exceed the lifting capacity of both male and female operators.

When determining the AL and MPL consideration is given to the horizontal position of the load in relation to the body at the point of lifting, the vertical position of the load at the point of lifting, the vertical displacement of the load during the lift and the frequency of the lifting operation.

The working environment

Space constraints

The importance of maintaining good postures during the lifting and moving of a load has already been highlighted. However, the working environment may interfere with the postures adopted by the operator which can place them at increased risk when they attempt to move an object.

Minimal space either around or above the operator will make the movement of an object difficult. Operators should be able to adopt reasonably upright postures when lifting and carrying objects. There should therefore be sufficient floor space. Ideally, they should not have to manoeuvre through narrow passageways. High standards of housekeeping should minimise the possibility of tripping or falling.

Overhead clearance should be provided in any workplace design to remove the need for operators to stoop when moving objects. This is particularly relevant when operators are working in storage areas with racks at different levels – they will have to stoop under upper racks when retrieving items from the floor level storage area. If objects on the upper racks are not intended to be moved by hand, and are in fact lifted down on pallets by forklift, then these racks should be moved up out of the way. When setting the heights of the upper levels, consideration should be given to whether operators in that area wear head protection. The use of hard hats will, of course, increase the overhead clearance required.

If it is not possible to increase the height of the upper racking, operators should be encouraged to slide the load forwards until it clears the upper rack before it is picked up.

Variation in levels

Steps and slopes make the handling task more complex – particularly if the operator needs a free hand to steady themselves during the manoeuvre. The use of steps should be of particular concern.

Slopes should be as gentle as possible and steps should be well maintained and illuminated. It may be beneficial to provide operators with a small ramp to assist in accessing different levels (eg over a kerb). However, if the ramp has to be removed after its use it should be treated as an additional load and its weight and point of storage should be taken into account during assessments.

If working between different worksurfaces such as a conveyor and a workbench, it is recommended that they are of a uniform level to remove the need to lift or lower.

Floors

The floor surface will have an obvious impact on the safe movement of a load and it should therefore be kept level and in good repair. This is equally important whether working inside or outside. Torn lino or carpet, for example, can provide a tripping hazard, while potholes and other rough surfaces increase the likelihood that the operator will have an accident.

Spillages such as oil, food, paint or water should be cleaned up as soon as possible. If the floor becomes wet on a regular basis slip-resistant surfacing should be considered.

Environmental factors

The temperature in the work area will have an effect on the handling operation. If the operator becomes hot the hands may become sweaty and objects could slip out of them. If they become cold, they may start to lose some of their dexterity. This is particularly the case for those who work outdoors or in cold stores. The provision of suitable clothing may reduce the effects of the cold but unless it fits properly it may hinder movement and make gripping more difficult.

Poor ventilation will have a tiring effect on the operator.

Sudden gusts of wind may throw the operator off balance when moving a large load (eg a sheet of glass or plasterboard). Action should be taken to reduce the possibility of this occurring – for example, taking a different route, using more than one person to move the load, using mechanical assistance, or re-packing the load into a more suitable container.

Poor lighting will increase the likelihood that the operator will trip while carrying the load and will make it more difficult to locate and deposit the load quickly. By extending the time it takes to deposit the load, the fatiguing rate of the operation is increased.

The location of lighting is important. The ceiling may not always be the most appropriate position for the light source – for example, when using racking systems upper racks may cast shadows over lower levels. Wall lights may be more suitable.

Further environmental factors such as excessive noise, rain, poor air quality or dust may encourage operators to hurry their handling operation, which will increase the risks associated with moving the load. These other 'incidental' factors should therefore be addressed.

Individual capability

Sometimes it is the personal characteristics of the operator which increase the risk of the handling operation and these should be taken into account when assigning individuals to the task of lifting and moving loads. Some people may prove to be unsuitable for a manual handling operation.

Bearing in mind the high recurrence rate for back trouble, it would be advisable to give careful consideration to operators who are known to have problems with their backs before asking them to become involved in a handling activity. This is also the case for those who have suffered hernias, ULDs or other health problems.

Gender

The different capabilities of men and women can be a hotly contested issue. There is, generally speaking, a difference between men and women in terms of lifting strength, with men having a greater lifting capacity than women. However, there is some overlap with some women coping with a heavy lifting situation better than men – this will mainly depend on factors such as age, fitness, health, build, training and experience. For example, most female nurses who have been trained in the movement of patients will lift and move a load more safely than most male city executives who have sat behind a desk for 25 years lifting no more than a briefcase.

The HSE recommends that its numerical guidance aimed at male operators should be reduced by about one third to provide the same degree of protection to approximately 95 per cent of the female operators who work in an area. The HSE guidance (1998) provides specific numerical guidance for female manual handlers.

The effects of pregnancy and childbirth should also be considered. Once an employee becomes pregnant, the hormonal changes her body experiences will affect her ligaments, making her more susceptible to injury. In addition, as the pregnancy progresses she may start to experience postural problems which will put her at an increased risk of injury. (Similar problems may be associated with obese employees.)

Manual handling during pregnancy combined with long periods of standing and/or walking is now considered to have significant implications for the health of the woman and the foetus.

Once a woman informs her employer that she is pregnant, the risks to the health and safety of both her and her unborn baby must be assessed. To ensure that employers comply with their duties under the Management of Health and Safety at Work Regulations 1999 they should have a plan on how to respond to the pregnancy. This plan could include:

- reassessment of the task to determine whether improvements could be made;
- training to recognise changes in capabilities as the pregnancy progresses;
- possible job-sharing opportunities, relocation to another area or suspension on full pay if an appropriate alternative cannot be found;
- communication with the GP; and
- monitoring of the employee before and after the birth.

Consideration should also be given to women returning to work after childbirth.

Age

Age is also an issue that may engender some debate. It is believed that an individual reaches their peak in terms of physical ability in their early 20s, followed by a decline during their 40s which becomes more marked beyond this age.

Many organisations which have an 'older' workforce should not be unduly concerned that their employees may appear to be more at risk than a 'younger' workforce. Older operators have the benefits of experience and maturity which should reduce the risks and may outweigh the benefits to be gained by having a younger, less experienced and more immature workforce. However, older operators should be made aware that although they may believe themselves to be fit and healthy, they should appreciate that they may become injured more easily and may take longer to recover from injury.

Disability

The Disability Discrimination Act 1995 places a duty on employers to make reasonable adjustments to the workplace or working arrangements if a disabled person is at a substantial disadvantage when compared with their non-disabled colleagues. This might result in the employer providing the disabled person with a handling aid or changing the size, weight and number of loads handled.

'Positive discrimination'

'Positive discrimination' can, and should, play a part in reducing the risks faced by a particular operator. Height and build are obvious factors which require such consideration. Quite often an operator will be working above head height as they reach to a rack or hopper simply because they are not very tall. Using a taller operator will reduce the risks of that particular handling operation and will not cost the organisation anything more than the time it takes to locate another operator.

The general rule, however, when considering the risks of a handling operation should be that if a reasonably fit and healthy individual cannot do the job comfortably and safely then it is unacceptable.

IN SUMMARY

The task

- The operator should adopt the correct posture before attempting to move an object.
- Loads should not be lifted or held at a distance from the body.
- Heavy, difficult to handle objects should be stored between knuckle and shoulder height with lighter, easier to move objects stored above or below this range.
- Shelves or racking should be marked to indicate which objects are to be stored at which levels so as to accommodate the weight-shape consideration.
- Platforms, duckboards or steps should be made available if objects are to be moved at higher levels. None of these should constitute a tripping hazard.
- Stacking heights should be controlled so that operators do not have to work at high levels.
- Lifting from floor level should be avoided. If possible, objects should be raised off floor level.
- If lifting from floor level, the load should not have to be raised above waist height in one move.
- Walking distances should be kept to a minimum by reorganising the layout of the work area.
- When pushing or pulling an object, the hands should not have to move outside the zone between knuckle and shoulder height.
- Unpredictable, sudden movements of the load should be controlled as far as possible.
- Additional care should be taken if the operator is handling a load while seated. The weights which are handled should be significantly reduced and the loads should not be placed on floor level or be at a distance from the operator.
- When two or more operators are involved in a team lift, they should be of equal size and ability, there should be sufficient space to move easily and care should be taken that each team member does not impede the view of the other(s).
- Rotation and regular rest breaks should be used to combat the stresses of handling.

The load

- Reducing the size of the load will make it easier to move.
- Consideration should be given to the fact that a bulky item may be very difficult to pick up (even though it may be fairly light).
- A load may be so large that it impedes the operator's view when it is picked up or drags on the floor during movement. This will increase the risk of the handling operation.
- The distribution of the weight of the load is as important as the weight itself. If the weight is not evenly spread, the heaviest side should be held close to the body.
- Objects lacking rigidity should be moved inside a container wherever possible.
- Free-moving objects should be packed tightly within their container using materials such as polystyrene or bubble-wrap.
- Liquids and powders should be filled as close to the top of the container as possible to limit free movement during handling.

- Objects should be clearly marked to indicate their weight, distribution of weight and any additional handling guidance.
- Markings on loads should be in a form that is easy to understand – colour coding or symbols may be more appropriate than written instructions.
- Numerical guidelines, such as those provided by the HSE (1998), should be used in determining the level of risk associated with the movement of any object.

The working environment
- Operators should have sufficient space around and above them to allow the adoption of suitable postures when moving loads.
- The floor should be as flat and smooth as possible with minimal numbers of steps and slopes.
- Temperature, humidity, ventilation and lighting levels should be assessed to ensure that they do not make the handling of loads more difficult.

Individual capability
- As a general rule, males have a greater lifting capacity than females. However, a degree of overlap exists and this will be influenced by health, strength, fitness, training and experience.
- Pregnancy influences a female's ability to lift and move objects safely.
- Individuals are at their peak in terms of lifting and moving objects in their 20s and start their decline in their 40s. This should be viewed in parallel with the benefits of experience and maturity, which are usually lacking in younger operators.

Training and awareness programmes

Training and information programmes should be used as a means of 'protecting' the individual once in a situation where they must manually handle loads. The programmes should provide operators with an understanding of good lifting techniques, as well as increasing their awareness of the need to develop a sensible approach to the whole working environment, not just the actual lifting of the load. Only an appropriately trained coach should present training sessions.

One of the most common statements made by operators is that they have been working in a particular manner for a long time and so far they have not experienced any problems. They assume that because they have remained relatively healthy to this point that nothing will happen to them in the future. Awareness programmes need to convince operators that most handling injuries result from the cumulative effect of years of abuse rather than from a one-off lifting situation. They need to be made aware of the fact that it is important to use proper lifting techniques each time they pick up an object – and not just at work.

Operators need to have an understanding of how to recognise the hazardous aspects of a handling situation so that they can take action to reduce the risks. They need to have a working knowledge of how to determine whether the load can be managed unaided. When faced with an unfamiliar load, operators should be advised to:

- rock the load from side to side to assess its likely weight before they attempt to lift it so that they can decide whether they are able to pick it up;
- rock the object on the edge of a worksurface to find out whether the weight is evenly distributed;
- test a box or barrel first to see whether it is empty; and
- lift the unfamiliar weight slowly from the surface, stopping before they clear the surface if it is apparent that the weight is too much for them to bear alone.

If operators are given any handling aids or protective equipment they need to have training in their use. Training programmes should also discuss the features of the workplace which have been put in place to increase levels of safety. It should be made clear to operators that good housekeeping – which is within their control – has an effect on the level of risk in the workplace.

Those who work in environments which are not viewed as 'typical' handling environments, such as an office, should be made aware of the fact that their ordinary clothes may not be suitable for the movement of loads. For example, a slim-fitting skirt may prevent the individual from adopting appropriate lifting postures. If there is a significant amount of lifting involved in a particular work area, the provision of protective shoes may be appropriate.

The operators should be made aware of how their individual characteristics and state of health can influence the level of risk of the handling operation (they should therefore act accordingly if they find that they are pregnant or unwell).

Finally, to ensure that operators lift objects safely it is essential that they receive training in good handling techniques – bearing in mind that good lifting techniques alone will not overcome the problems presented by a poor working environment.

If training is to be effective, operators should learn by doing, not watching. It is said that a person remembers:
- 10 per cent of what is read;
- 20 per cent of what is heard;
- 30 per cent of what is seen;
- 50 per cent of what is seen and heard; and
- 70 per cent of what they do themselves.

At the start of the training programme, the operators should be made aware of the purpose of the exercise. Once they have received an explanation of lifting and handling in general they should be asked if they understand the basic principles before being asked to lift any object. Once operators start to lift and move objects they should receive feedback on their performance to enable them to correct any mistakes before bad habits are adopted.

Good handling techniques

The movement of the load should be thought of as the 'performance' aspect of good handling techniques. However, the operator should consider planning and preparation before starting to handle the load:
- **Planning** – before handling the load the operator should decide whether the load actually needs to

be lifted or moved. If so, they should not assume that only their hands will do the job. Instead, they should consider whether there is any equipment available which will help. If not, they should consider whether another person should help. They then need to determine whether there is enough room to move the load safely. Once they have planned how the load will be moved they should prepare for the lift.

- **Preparation** – the operator needs to ensure that they have suitable clothing and footwear which will not hinder them during the movement. They need to have an idea of the weight of the load and how the weight is distributed. They will also need to determine whether the edges are safe to hold and whether they can obtain a secure grip.

This procedure for planning and preparation may appear to be very time-consuming but an experienced operator will be able to run through the planning and preparation list mentally in a matter of seconds before they attempt to move an object. In most repeated lifting situations, all the planning and preparation will have been carried out during the job design phase so will not have to be done by the operator at all.

Before moving on to the 'action' phase of moving a load, the operator needs to develop appropriate performance skills. These skills require practice so that they eventually become a natural part of the handling process. It is recommended that small, compact loads of approximately 5 kg are used for practice purposes in controlled manual handling training sessions.

Operators should be trained in the basics of a lifting operation, starting with a squat lift where the load is raised from the floor:

- The operator should stand close to the load with the feet hip-width apart.
- One foot should be placed slightly in front of the other and to the side of the load if possible (this will give a stable base from which to lift an object).
- The knees should be bent and the back kept straight. The back does not have to be bolt upright – the operator can lean over the load slightly. However, the operator should not kneel on the floor or have a deep kneebend because this will make it more difficult to stand up.
- The load should be grasped firmly using the entire hand, not just the fingers.
- Having grasped the load the operator should look up, which not only lets them see where they are going but also helps to straighten the back.
- The load should be lifted smoothly using the leg muscles to stand up.
- The load should be kept close to the body. The heaviest side should be held closest to the body.
- Once standing upright, the feet should be used to turn around. Operators should not twist at the waist to deposit the load onto another surface.
- The load should be deposited slowly using the same sequence in reverse. Precise location of the load can be accomplished after it has been put down.

Once the squat lift has been learnt, operators can apply the same technique to moving an object from a raised surface. They will see that they follow the same principles, despite the fact that the load is being moved from a higher level.

If the operator is required to push or pull a load (eg a trolley or clothes rail) other considerations should be incorporated in the training session, including:

- The operator should keep the load's centre of gravity aligned with the body's own centre of gravity – directly in front of the body is best.
- Turning should be accomplished using the feet, not the upper body.
- All effort should be channelled through the legs. The operator should be encouraged to grip the load firmly and propel themselves forwards or backwards using their leg muscles.
- Sudden or jerky movements should be avoided.
- The operator should be as close to the load as possible when moving it. The arms should not be stretched out in front to apply pressure when pushing. Ideally, the elbows should be bent and close to the bodyside.

Once operators have become used to the movement of small, compact loads using the appropriate lifting techniques they should progress to moving more difficult loads. At all times they should be aware that they must recognise their own limitations and stop if any load is too much for them to cope with.

Once operators have accomplished the movement of the heavier, more difficult loads they should move on to the movement of loads they can expect to deal with in the 'real' working situation (there is no point in only learning to move a 5 kg compact box when they are expected to move gas cylinders or sides of beef once back at work). The training programme should identify a range of loads the operators will be expected to move, and provide, if necessary, specific tuition on particular loads. Drums, barrels, cylinders, sacks and other problematic loads should be included in handling programmes for those operators who are expected to move such items by hand. A special lifting technique may need to be adopted for each item and operators will need to have the opportunity to practise under controlled conditions.

Training programmes should also provide an opportunity for operators to lift in teams so that they can come to terms with the additional problems associated with team lifting.

Problems preventing introduction of good lifting techniques

On returning to the workplace following a training session, there is no guarantee that operators will continue to use their newly learnt lifting techniques. Managers and supervisors need to be aware of the reasons why people may revert to their old habits so that they can discourage them.

Some operators try to prove how strong and fit they are and develop foolhardy methods of work when trying to move twice as much as others. They should be made aware of the problems they could cause if they do not develop a more mature attitude. Other operators who are unable to cope with the movement of a heavy load may attempt to 'soldier on' rather than seek help because they do not want to draw attention to themselves. These individuals should be made aware that they are also putting themselves at risk and should be more confident about asking for assistance. To facilitate this attitude change there will need to be a suitable response from others in the workplace when help is sought. Some operators will not use the proper handling techniques because they mistakenly believe that it is quicker the 'old' way. Finally, the main reason why manual handling training may become virtually redundant once the operator returns to work is that it is very difficult to change long established habits. It is important to convince operators that although they may have avoided injury so far, long term abuse will probably catch up with them in the end.

If operators are to change their old habits they must be supported by appropriate supervision which encourages the use of proper handling methods and discourages inappropriate habits. Supervisors should lead by example. They should not ask someone to move something they would not move themselves, and if they do lift an object they should make sure that they use the relevant techniques.

Back belts

Some individuals believe that using a back belt while manually handling a load will reduce, if not eliminate, the problems associated with its movement. Other operators are under the impression that they cannot take a claim out against their employer in the event of an injury if they were not wearing such a belt.

The use of back belts should be considered very carefully. It should be understood that it is unlikely that operators will avoid injury through wearing a belt, and in fact, some operators may develop a false sense of security as a result of wearing one. If belts are provided, operators should be given training in their appropriate use and supervisors should ensure that they are worn properly at all appropriate times.

If back belts are already in use, it is not recommended that operators are prevented from wearing them. However, operators should not be required to wear a belt unless they wish to do so.

IN SUMMARY

- All those involved in the moving of loads should be given manual handling training. (This includes office personnel who sometimes are under the mistaken impression that they do not lift things at work.)
- Operators should learn that they have to plan and prepare for moving an object before they attempt to move it. This includes ensuring that they have enough room to move around, considering whether they need assistance and assessing the distribution of the load's weight. Operators should be advised to test all unknown objects before attempting to lift them.
- Training should be given in the use of handling aids and PPE as well as manual handling techniques.
- On return to the working environment following a training programme, operators should be supported by appropriate supervision which encourages the continued use of safe handling techniques.

REFERENCES AND FURTHER READING

Ayoub M M, 1989, *Manual materials handling: design and injury control through ergonomics*, Taylor & Francis, London.

Ayoub M M and Mital A, 1989, *Manual materials handling*, Taylor & Francis, London.

Birnbaum R, Cockroft A and Richardson B, 1993, *Safer handling of loads at work – a practical ergonomic guide* (2nd edition), Institute of Ergonomics, Nottingham.

Cremer R and Snel J, 1994, *Work and ageing: a European perspective*, Taylor & Francis, London.

De Beeck R O and Hermans V, 2000, *Research on work-related low back disorders*, European Agency for Safety and Health at Work, Bilbao.

Health and Safety Executive, 1994, *Manual handling in drinks delivery*, HSG119, HSE Books, Sudbury.

Health and Safety Executive, 1994, *Manual handling: solutions you can handle*, HSG115, HSE Books, Sudbury.

Health and Safety Executive, 1998, *Manual Handling Operations Regulations 1992: guidance on Regulations*, L23, HSE Books, Sudbury.

Jenkins D B, 1998, *Hollinshead's Functional anatomy of the limbs and back* (7th edition), W B Saunders, St Louis MO.

Kroemer K and Grandjean E, 1997, *Fitting the task to the human* (5th edition), Taylor & Francis, London.

Mital A, Nicholson A S and Ayoub M M, 1993, *A guide to manual materials handling*, Taylor & Francis, London.

National Institute for Occupational Safety and Health, 1981, *Work practices guide for manual lifting*, NIOSH, Cincinnati OH.

Pheasant S and Stubbs D, 1995, *Lifting and handling: an ergonomic approach*, National Back Pain Association, London.

Troup J D G and Edwards F C, 1985, *Manual handling: a review paper*, HMSO, London.

UPPER LIMB DISORDERS

Introduction

As long ago as 1713 Bernadino Ramazzini identified a condition in scribes which has since become known as repetitive strain injury. Over time it became evident that it was not just repetitive work that was responsible for the range of symptoms and disorders experienced by people in their hands, arms, shoulders and neck and the condition was renamed upper limb disorder (ULD). This term offers a more acceptable description of what the affected person may experience.

Despite the fact that in the UK the term ULD is used fairly widely, many people still debate the accuracy of a term that is not universally accepted. For example, in Japan the same condition is referred to as occupational cerviobrachial disorder (OCD) and in the US the term cumulative trauma disorder (CTD) is used. Sometimes the argument is put forward that if there is not even agreement on the name that should be used, how can the causes of the condition in the workplace be tackled? Others argue that it does not really matter what the condition is called – as long as it is recognised that the person has been injured in some way by a process or arrangement at work and that steps should be taken to tackle the issue.

Although some dispute that there are such things as work-related ULDs, this chapter takes the stance that ULDs do exist and that they may be caused, or exacerbated, by elements in the workplace.

Having drawn attention to the work-related causes of ULDs, it should be understood that these are by no means the only causes of these conditions. It is now generally accepted that certain hobbies, domestic activities and medical conditions can play a part in the development of a ULD. Therefore, when an individual reports pain or discomfort in the upper limbs, a holistic approach should be taken to identify the possible cause(s). It should not be assumed that work alone is to blame.

Postures of the upper limbs

Prior to considering ULDs in any detail, it is advisable to explore the postures adopted by the upper limbs during the course of a day. An appreciation of the range of postures used will assist in understanding why certain wrist or arm positions are more likely to be stressful for the limbs than others. Each major joint in the upper limbs can display a wide range of movements. However, any movement that pushes the joint away from a neutral or natural position can be harmful, particularly if carried out repeatedly or for extended periods of time.

When describing the work being carried out by the limb, it is important that an accurate description is given of its position or direction of movement. For example, is the palm facing up or down; is the arm alongside the body or stretched out to the side? Because of the importance of describing limb positions accurately, a specific vocabulary has been developed which describes the positions adopted by the whole arm or just part of it. Some of the more commonly adopted limb positions are shown on the next page and a description of the terms used in the illustration follows.

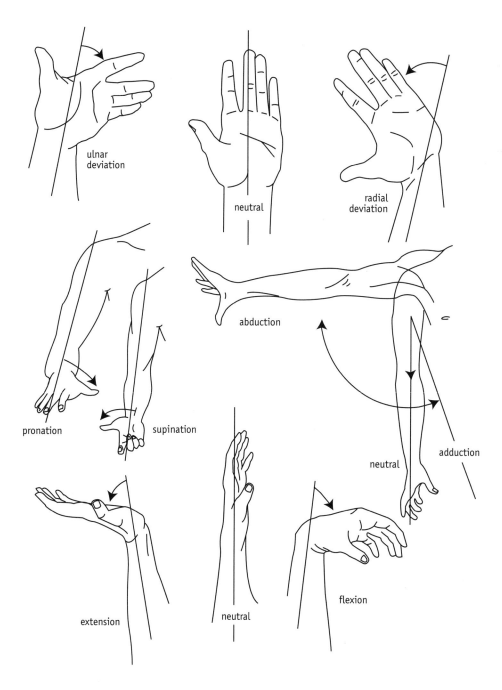

Commonly adopted upper limb positions

Reproduced from 'Cumulative trauma disorders: a manual for musculoskeletal diseases of the upper limbs' (Putz-Anderson 1988)

- **Abduction** – moving the arm outward and away from the side of the body.
- **Adduction** – moving the arm across the front of the body.
- **Pronation** – rotation of the forearm which results in the palm of the hand facing down.
- **Supination** – rotation of the forearm which results in the palm of the hand facing up.
- **Radial deviation** – bending the hand at the wrist in the direction of the thumb.
- **Ulnar deviation** – bending the hand at the wrist in the direction of the little finger.
- **Wrist flexion** – bending the hand down at the wrist.
- **Wrist extension** – bending the hand up at the wrist.

Types of ULD

It is generally accepted that ULDs can affect any area of the upper limbs from the fingertips to the neck. Specifically named disorders usually identify which part of the body is affected and what type of symptoms the person is suffering. The following section outlines some of the more common disorders that can be experienced.

Hand and wrist area

Tenosynovitis

This term has probably been used more often and more inaccurately than any other condition identified as a ULD. It was regularly used in the late 1980s and early 1990s as an umbrella term to denote that a work-related upper limb disorder had occurred. Tenosynovitis is, in fact, a specific ULD. It arises from inflammation in the lining of the synovial sheath of the tendons, most commonly in the hand and wrist. The tendons and their sheaths are shown below.

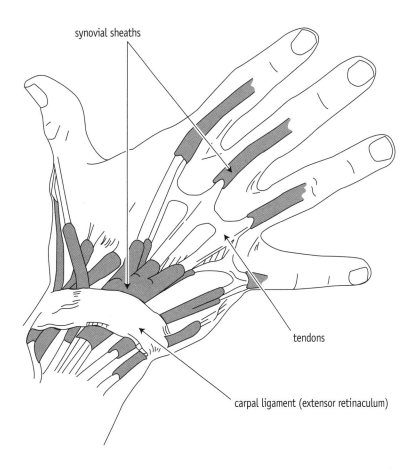

Tendons and sheaths in the hand

Reproduced from 'Cumulative trauma disorders: a manual for musculoskeletal diseases of the upper limbs' (Putz-Anderson 1988)

The tendons normally move freely inside the sheath which is lubricated and which offers protection as the tendon passes under ligaments or around corners such as when the wrist is bent. Repetitive work is often cited as a cause of tenosynovitis. However, it is generally accepted that a traumatic injury, such as a heavy fall or blow to the wrist, may predispose an individual to developing tenosynovitis, particularly if they are employed to carry out an activity where the work is manually intensive. Other work-related factors which are closely associated with the development of this condition are forceful gripping, usually when carried out repetitively, and working with the hands away from the neutral position (illustrated on page 136). Those suffering from tenosynovitis may experience symptoms such as an aching in the wrist, a weakness of grip, swelling (or even just the sensation of swelling) and sometimes crepitus which produces a crackling sound similar to the noise made when someone walks over dry snow.

A slightly different condition called stenosing tenosynovitis is caused if the movement of the tendon inside the sheath is impaired. If the abductor pollicis longus and extensor pollicis brevis tendons are involved the condition is referred to as De Quervain's Disease. These tendons can be easily seen on the hand at the base of the thumb. They form the outer wall of the 'anatomical snuffbox' – the small indent in the surface of the hand just above the wrist area on the side of the thumb. Prolonged exertion, repeated strain or unaccustomed work have been implicated as precipitating factors in this particular condition. Operations which involve deviation of the wrist combined with forceful grasping, particularly of larger objects where the thumb is moved outward away from the hand to wrap around the side of the object, are also considered to be more likely to contribute to the development of this condition.

If the tendons of the finger flexors are affected the individual will be diagnosed as suffering from trigger finger. The flexors are the tendons which allow the fingers to be pulled in towards the palm, such as when making a fist. This condition usually results from over-use of the fingers and is often associated with the repeated or extended use of tools, particularly when the handle or trigger has hard or sharp edges. Sufferers may find that they can only make jerky movements of their fingers.

Carpal tunnel syndrome

The tendons responsible for flexing the fingers, as well as the median nerve and blood vessels, pass from the forearm into the hand through the carpal tunnel under the carpal ligament (illustrated on the previous page). The median nerve, which is shown opposite, is responsible for the sensation in the thumb, the majority of the palm, the index finger, middle finger and part of the ring finger.

Carpal tunnel syndrome is caused by compression of the median nerve. This compression can result from irritation and swelling within the confines of the carpal tunnel such as occurs in tenosynovitis. It can also result from the repeated deviation of the wrist, repeated forceful gripping and the vibration connected to tool use. The syndrome has also been associated with pregnancy, the menopause, a local trauma such as a fracture, diabetes and rheumatoid arthritis.

Typical symptoms of carpal tunnel syndrome are numbness and tingling, usually in the areas of the hand connected to the median nerve. Symptoms are known to be so severe at night that they can wake the sufferer. There may be a loss of sensation in the hand which can make it difficult for the sufferer to pick up small objects, giving the hand a rather clumsy feel.

UPPER LIMB DISORDERS

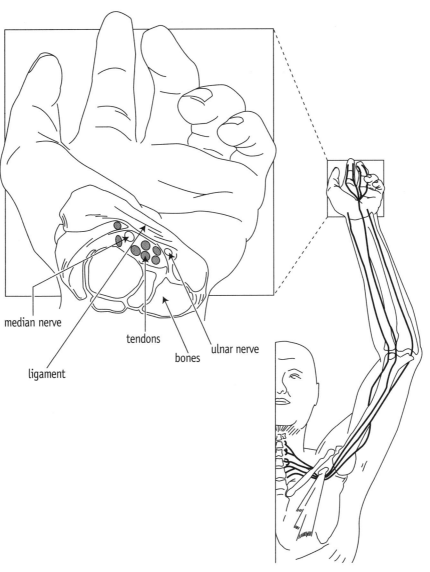

Cross-section of the wrist

Reproduced from 'Cumulative trauma disorders: a manual for musculoskeletal diseases of the upper limbs' (Putz-Anderson 1988)

Dupuytren's Contracture

This condition is caused by a shortening and thickening of the fibrous fan on the palm of the hand known as the palmar fascia. The shortening and thickening causes a gradual and permanent bending of the fingers unless surgical treatment is sought. The most common digits to bend are the little and ring fingers.

Repeated minor trauma to the palm of the hand (such as occurs when a carpenter hits a chisel end with the palm) and the use of vibratory tools may cause the condition. Specific injury, such as a laceration to the hand, may also contribute.

Dupuytren's Contracture has been shown to be congenital and may not therefore have a work-related component in some sufferers.

Ganglion

A ganglion is a fluid-filled swelling, usually found on the back of the wrist or hand. It is generally about the size of a large pea but can be bigger. Although ganglions are not painful, and some would suggest that they are not work-related, it is usually considered that a high incidence in a particular workplace is often a sign that the operators are exposed to inappropriate working conditions (eg an excessive work rate) or that they are adopting irregular upper limb postures while working.

Lower and upper arm

Epicondylitis

There are two types of epicondylitis – commonly known as tennis elbow and golfer's elbow. Tennis elbow is more accurately referred to as lateral epicondylitis and affects the side of the elbow which is in line with the thumb when the arm is held out in front of the body with the palm facing upward. Golfer's elbow affects the opposite side of the elbow.

Lateral epicondylitis means that the area around the epicondyle – the bony bump on the outside of the elbow – is inflamed. The epicondyle is the point where the muscles responsible for extending the wrist and fingers originate. Over-use of these muscles can cause pain as the epicondyle becomes tender and swollen.

Forceful movements that bring the hand upwards at the wrist (eg when throwing an object) or repeated rotation of the forearm (eg when using a screwdriver) are considered to be likely to contribute to the development of lateral epicondylitis. Heavy lifting, particularly when the back of the hand is facing upward (eg when lifting a brick), is also considered to be associated with the development of this condition.

Golfer's elbow, or medial epicondylitis, is very similar to lateral epicondylitis but is not as common. It is an irritation at the point of origin of the flexor muscles on the inside of the elbow and is usually caused by repeated flexion of the wrist, ie where the hand is bent downwards at the wrist, and rotation of the forearm.

Tendinitis

Tendinitis is a condition which can affect any tendon in the body. Humeral tendinitis, or rotator cuff tendinitis, is an inflammatory condition which affects the tendons of the muscles responsible for the rotation of the arm at the shoulder and the movement of the arm away from the body. Other surrounding soft tissue, such as the subacromial bursa, is also affected. The subacromial bursa is a cushion which protects the tendons from the bony ridge that overhangs them during movement. This bony ridge is the acromian.

Shoulder problems, such as humeral tendinitis, are generally found in operators who work with their arms raised for extended periods, or those who raise their arms repeatedly.

'Frozen shoulder'

Continued overhead work, where the arms are repeatedly raised or held in this position for an extended period, can lead to thickening of both the tendons and bursa. This results in the condition known as frozen shoulder which causes the sufferer to experience pain and impaired function. However, many cases of frozen shoulder have been shown to have no connection with work activities.

Thoracic outlet syndrome

The condition involving the nerves and adjacent blood vessels in the shoulder and upper arm is known as thoracic outlet syndrome. It is a general term used to indicate that the nerves and blood vessels between the neck and shoulder have been compressed. The individual will experience similar symptoms to those of carpal tunnel syndrome, such as numbness in the fingers.

Work involving frequent reaching above the shoulder, or activities which require the shoulders to be pulled backwards and downwards (eg when carrying a stretcher) are known to contribute to the development of this condition. People who repeatedly carry weights directly on their shoulders (eg brick-layers supporting a hod) or who suspend weights from their shoulders (eg postal workers) are more likely to experience this condition.

Other conditions

Cervical spondylosis

Cervical spondylosis is a common condition of the neck and spine. It is an inflammation of the synovial joints of the neck and is often associated with age. It has also been associated with those who habitually carry heavy loads on their shoulders (eg bricklayers using hods).

Osteoarthritis

Osteoarthritis is a degenerative condition more accurately known as osteoarthrosis. It is caused by general wear and tear which affects the articular cartilages of the synovial joints. Localised osteoarthritis can occur when a particular joint has been subjected to long term stress (eg the elbow joints of slaughterhouse workers involved in pulling skins off animals).

Causes of ULDs

It is widely accepted that most ULDs are caused by easily identified aspects of the workplace. Numerous sectors (commercial, service and industrial) and occupations are closely associated with high rates of ULDs among the workforce. In most instances where an individual develops a ULD it is a result of a combination of factors rather than exposure to one single factor.

There are various 'main' causes of ULDs and certain additional 'aggravating' factors. More often than not, these act together to the detriment of the individual.

Main causes of ULDs

The main causes of ULDs include repetition, poor posture, force and static work. However, it should always be borne in mind that just because an activity includes one or more of these elements and therefore has the potential to cause harm, it is not inevitable that an operator will develop a ULD. The outcome is mediated to an extent by contributory or aggravating factors – discussed later in this chapter.

Repetition

Any job that requires an operator to carry out the same operation or sequence of operations repeatedly within a short timescale should be considered to put them at risk. Muscles that have to work rapidly develop less tension than muscles that can work more slowly. As a consequence, more effort is required to complete an operation carried out at a fast rate. As a result of applying more effort, the operator requires a longer period of time for recovery, which work-rest schedules rarely take into account. It was thought at one time that repetitive work would only cause a ULD if it was combined with excessive force. However, it has since been accepted that high rates of repetition are sufficiently stressful to promote the development of a ULD without excessive force being present. People who work in manufacturing and assembly facilities are often involved in activities that could be described as short cycle and highly repetitive. The tasks are considered to be short cycle because the sequence of movements is of a limited duration and the sequence is repeated at the end of each cycle.

Putz-Anderson (1988) classified tasks on the basis of the cycle time and the percentage of cycle time performing the same fundamental cycle. Jobs can be categorised as low repetitive if the cycle time is more than 30 seconds, or if under 50 per cent of the cycle time involves performing the same kind of fundamental cycle. Jobs can be categorised as high repetitive if the cycle time is less than 30 seconds, or if more than 50 per cent of the cycle time involves performing the same kind of fundamental cycle. An example of where the same fundamental cycle is repeated can be seen where operators remove food products from conveyor belts and pack them into outer cases. It may take operators 40 seconds to fill one outer case which might suggest that the task is not highly repetitive. However, they place 20 products in the outer case at a rate of one product every two seconds (ie the fundamental cycle is repeated every two seconds) which means that the task would be classified as high repetitive. Operators who carry out a task that falls within the high repetitive category are considered to be at a greater risk of developing a ULD.

Other factors within a facility, apart from the speed of the line or general production rate, may influence the repetitiveness of the operation. For example, a piece-rate system of pay or a productivity target may encourage operators to work at a rate that could move them out of the low repetitive bracket into high repetitive. It is also possible that an operator might unintentionally make their own task high repetitive. For example, an operator may be given freedom over how an assembly operation is completed. The operator may be required to make a number of products within a specified time. The product could be made up of several individual parts, each of which is attached in a different way using a variety of tools and manual effort. In a bid to speed up the rate of production, the operator may decide to abandon the method of work where one complete unit at a time is assembled and opt instead to repeat the same operation on a number of units, adding the same part repeatedly to each successive unit. The operator would then move on to attach the second part in the same way, resulting in a second highly repetitive operation (and so on). The 'redesign' of the operation results in a series of highly repetitive sub-routines which are likely to accelerate upper limb fatigue and enhance the development of a ULD.

There are various means available to combat the repetitiveness of an operation, the most obvious being a reduction of the speed at which the operator works. Alternatively, more regular breaks in activity could be provided – as formal rest breaks or job rotation. If rest breaks are to be effective, they should be taken at regular intervals distributed evenly throughout the shift. The success of rest breaks lies in their timing, not in their length. The intention should be to allow operators to *rest* before they reach a point where they are tired. Once an operator reaches this point, a rest break becomes a *recovery* period only. Short, frequent breaks are better than long, infrequent ones.

If job rotation is to be effective, each successive task should be qualitatively different from the preceding one. Simply moving an operator to a different location to carry out a second similar task will not provide the intended relief. This form of rotation is commonly found in manufacturing and assembly operations where the programme of rotation dictates that the operators follow the same production path as the product which often results in them moving through a series of similar tasks. For rotation to work properly, an analysis of the activities available to the operator should be assessed and an appropriate rotation programme developed. During the analysis, consideration should be given to the degree of repetition, the postures adopted and the movements involved. In assembly and manufacturing operations, it may be necessary for operators to follow a path where they miss out one or two tasks in the production sequence and then double back to these at a later point in the rotation schedule.

Posture

The posture adopted by the upper limbs while the operator is working plays a significant part in the development of ULDs. If an operator is required to over-extend a joint (such as when reaching forwards to pick an object off a conveyor belt) or reach upwards (such as when applying plaster to a ceiling), they may experience problems if the action is carried out repeatedly or for a sustained period of time. Movements of the wrist from side-to-side or up and down – which shift the wrist away from the neutral position – are also considered likely to cause a ULD if carried out rapidly or repeatedly or sustained for extended periods. Poor posture is a regularly occurring feature in many workplaces. For example, it is not uncommon to observe a keyboard operator, who has not adjusted their chair properly, working with their hands bent upwards or downwards at the wrist, depending on their sitting height in relation to the height of the keyboard. The sustained deviation of the wrist could lead to the development of discomfort in this area, particularly if regular breaks away from the workstation are not taken.

Ideally, people should be able to carry out their work with their upper arm hanging naturally at the side of the body with the forearm bent no higher than where it forms a 90° angle at the elbow. A straight line should run from the elbow, through the wrist and into the hand. This allows the person to work within their normal working area (the concept is discussed in the chapter entitled 'The design process' and relevant data are offered in the 'Anthropometrics' chapter). Deviation from this position increases the stresses placed on the upper limb and if such deviations are carried out repeatedly, or are sustained for an extended period, they can cause problems.

Particular postures of the upper limb that have been identified as being stressful (shown on page 136) include: ulnar deviation, radial deviation, extension, flexion and pinching (where the tip of the thumb and the tip of the index finger are brought together to hold something). The pinch grip is five times more stressful than the power grasp where the hand can be closed firmly around an object.

The posture of the whole arm is usually dictated by the workstation design and layout. If the working heights and reaching distances have not been designed to fall within the anthropometric parameters of the working population, then the operators have to overcome the shortfall by raising their shoulders and arms or reaching forwards with their arms. Operators working in these positions for any length of time may start to experience difficulties.

The deviations of the hand and wrist are usually dictated by the activity being carried out by the operator (eg tightening a screw, assembling a printer, sprinkling broccoli on top of a pizza). The hand is considered to be at its most powerful when it is in the neutral position or slightly extended. Grip strength may be reduced by up to 25 per cent if the hand is used when it is bent towards the little finger. If the hand is bent towards the thumb there may be a loss of grip strength of up to 20 per cent. This loss of strength emphasises the importance of allowing the operator to adopt a neutral posture while working. Working with irregular upper limb postures will ultimately be fatiguing for the operator and more likely to result in the development of a ULD. Consideration should therefore be given to finding a mechanical alternative to manual effort. For example, it is common to find the non-dominant hand being used as a clamp while the dominant hand uses a tool to tighten a nut or to cut through a material. Using a clamp would release the hand from the stressful task of gripping the item.

Workstations should be designed in a manner which reduces the need for the operator to work with their arms raised or outstretched. The chapters on 'The design process' and 'Anthropometrics' should provide sufficient guidelines for the development of appropriate workstations. Tools should also be redesigned or replaced in an attempt to eliminate deviation of the wrist. The chapter on 'Hand tool design and use' will assist with this issue. The task and the product being handled should be assessed to determine whether the task can be completed, or the product assembled, in an alternative manner which eliminates the need to adopt highly irregular postures.

Force

The degree of force used when carrying out a task is another factor associated with the development of ULDs. The amount of force that is used is determined, in part, by the tool being used or the objects with which the operator is working. For example, the weight, size and shape will influence how easy an object is to take hold of and this will dictate the degree of force required to grip it. In assembly operations, the ease with which parts fit together and whether they are off-specification will determine how much force an operator has to apply when completing the task. The pliable nature of some products influences the level of force required to complete a task. For example, a sewing machine operator may have to match up a number of individual pieces of fabric. If the fabric is tough and resists attempts to stretch it, the operator will have to pull forcefully to make the separate parts match up appropriately while being stitched.

Applying undesirable forces in a repetitive manner or for a prolonged period is likely to contribute to the development of ULDs. The level of force required for an operation should therefore be controlled as far as possible. This can be achieved through a number of measures, including:

- The effort required to close two independent units of a tool or a piece of equipment should be

reduced (eg when using spring-opening grips, the tension level of the spring should not require the operator to press firmly each time the grips are closed).
- Triggers on tools should offer the least amount of resistance possible while remaining at a level which is compatible with the avoidance of accidental activation.
- Cutting edges of tools such as shears, scissors and knives should be inspected regularly and kept sharp.
- Where possible, other sources of power should be applied to eliminate the need for manual effort (eg jigs and clamps should be used instead of requiring the operator to use a hand to clamp a part in place).
- The length of tool handles should be compatible with the degree of effort required to complete an operation. In some instances, a longer tool handle will afford greater leverage which will reduce the degree of effort required of the operator.
- Quality control measures should be put in place to ensure that only parts of appropriate specification are used in assembly areas.
- Operators should be trained not only in the method of completing a task but also in the amount of force appropriate to the operation (operators may consider that if they use more effort they will do a better job – however, they are simply more likely to tire at a faster rate).
- All equipment, tooling and mechanical aids should be included in a maintenance programme which schedules services on a regular basis (equipment that resists easy use by the operator will require an increase in applied effort).

Static work

Static muscle work occurs when muscles are tensed in a fixed position for an extended period of time. Discomfort and fatigue are typical features of static work. It can be observed in postures where the arm is held away from the body such as above the head or out sideways as the operator carries out an activity. During assessment of a task it is not uncommon for the observer to focus on the dynamic aspect of the operation (eg the hand rotating or polishing an object) and to overlook the non-movement involved where the arm is held in an inappropriate fixed position as the hand completes the work.

Muscles involved in static work need more than 12 times the length of the actual task duration to recover from the fatigue. Few work-rest schedules take this into account. Dynamic work, where the muscles contract and relax rhythmically, is less likely to result in fatigue and possible injury.

Alternative means should be developed to replace the need for the limb to be held in a fixed position. For example, instead of requiring the operator to fit an exhaust to the underbody of a car located overhead, the car could be rotated and presented on its side to the operator. This would mean that the operator could work in an area located directly in front of their body which would significantly reduce the extent of the static muscle work involved.

Other contributory factors

Temperature

Exposure to colder temperatures can increase the likelihood of an operator developing a ULD. Lower temperatures can be evident in an entire area (eg a chilled food storage area) or can be transferred to the operator through a point of contact (eg when holding a tool with a stainless steel handle – such handles are often resistant to warming up during a shift). Operators should be protected from the effects of working in a cold environment. In particular, the hands should be protected – for example, consideration should be given to the provision of appropriate gloves to operators who handle cold products such as chilled foods, or who work with air tools that produce a cold air stream during use. If the wrist and forearm area needs protection, gauntlets may need to be provided.

However, gloves or gauntlets should only be provided where it is not possible or practical to make a work area warmer through the use of heaters and so on. Prior to providing gloves, consideration should be given to the possible reduction in tactile capability and the increase in grip strength requirements that may result from their introduction.

Other controls include fitting air tools with a take-away hose which re-routes the stream of cold air, and covering tools with cold handles in a compliant material (eg rubber).

Various (perhaps less obvious) situations can involve the operator routinely coming into contact with cold surfaces. For example, an operator may lean on the side casing of equipment for support during a quality control task involving watching products passing by on a line. If the operator's bare skin, particularly the wrist and forearm area, comes into contact with the cold surface, a degree of heat exchange may occur.

Operators may not necessarily work in very cold environments but may be exposed to a sudden reduction in temperature at intervals throughout the shift. For example, operators working in a factory site may be located close to external doors that are left open for extended periods which can result in the internal temperature falling sharply. Others may work alongside chillers which may result in the room temperature falling each time the doors are opened. Protective steps may include:

- screening off the area to protect it from draughts;
- providing operators with thermal clothing and waterproofs;
- issuing gloves;
- providing heated pads to warm the hands; and
- supplying tools with heated handles.

Vibration

Vibration is known to be a possible cause of a number of hand injuries, including impaired blood circulation and damage to muscles and nerves. Operators who are exposed to vibration, such as when holding vibrating tools or workpieces, are at risk of developing a ULD. The level of risk is dependent on the vibration magnitude and how long the operator is exposed to it (ie the vibration dose). The level of risk is also affected by:

- **The degree of force employed** as the item held in the hand is gripped and pushed forwards – a tighter grip results in more vibration energy being transferred to the hand. This has implications for the design of the tool handle which will influence the degree of effort required to hold and use the tool. It also has implications for the use of gloves with vibrating tools and workpieces as this

sometimes impairs the dexterity of the hand, making it more difficult to hold the object and causing the operator to grip it even more firmly.
- **The exposure pattern of the individual** – this takes into account the frequency with which the operator carries out the task involving the vibrating items and the duration of the activity. It also takes account of any rest periods that might interrupt the work.
- **The area of exposure** – the vibration may be transferred to the whole hand or part of it. The area of exposure should be accurately identified so that the level of risk can be determined.
- **Individual issues** – some operators may be more susceptible to the effects of vibration than others (eg smokers may be more susceptible, given the effect of smoking on blood circulation).
- **Environmental conditions** – working in colder temperatures increases susceptibility.

If the risk of an operator developing a ULD is to be reduced, the level of vibration should be controlled. Options include:

- automating or mechanising the process so that vibrating tools do not need to be used;
- eliminating the need for the operator to hold vibrating workpieces by providing clamps and jigs;
- providing tools with handles that have been designed in a way which makes them easy to hold and reduces the forces required to hold them during use;
- ensuring that the operator has suitable training in the use of tools (how to hold and operate them correctly, including the required degree of force);
- using tools that are appropriate for the task;
- using tools that have been designed specifically for low vibration (eg fitted with anti-vibration mountings or vibration-isolation handles);
- maintaining tools and equipment so that they function properly and do not cause excessive, unnecessary vibration;
- ensuring that the operator's hands and body are kept warm;
- providing gloves that fit properly and which do not hinder the use of tools;
- limiting the operator's exposure time to the task involving the use of vibrating tools; and
- developing a programme for job rotation which limits the operator's exposure time to tasks involving vibration.

'Anti-vibration' gloves are not usually considered to be effective in reducing the amount of vibration reaching the hands. If such gloves are not an appropriate fit, they may even increase vibration transmission.

It may be beneficial to encourage operators who use vibrating tools to exercise and massage their fingers during rest periods as this will help blood circulation.

Organisational factors

As discussed above, the frequency with which an operator can take breaks away from work, whether through a rest break or job rotation schedule, will play a part in determining the likelihood of the development of a ULD. Clearly, overtime can play a contributory part in the development of an injury as it extends the length of time the operator is exposed to the injurious agents and reduces the overall amount of recovery time available in any one day. It may not be appropriate to extend the working day for those involved in high risk activities. It may be more appropriate for the overtime to be shared out between operators on an intermittent basis. (It is also worth noting that where excessive overtime is worked by any single operator, a reduction in their hourly productivity rates across the shift may result.) Generally speaking, excessive overtime may lead to increased absenteeism and accidents. Of course, an operator diagnosed as a ULD sufferer should not be permitted to work overtime unless it is specifically sanctioned by an appropriately qualified medical adviser.

Effective training and supervision play an important role in the control of the development of ULDs. It is not uncommon for on-the-job training to amount to little more than standing alongside an experienced operator and copying what they do (including any poor working practices).

It is important that all operators learn from the outset the safest and most efficient methods of work. Once they have learnt the techniques, they should be supported by appropriate supervision in the workplace. Managers, supervisors or team leaders may 'turn a blind eye' and take the attitude that as long as the job is done they do not care how it is done. Supervisors should actively encourage the adoption of appropriate working practices and discourage poor practices.

Peaks in the workload are also associated with the development of ULDs. It is considered that all operators require a period of time in which to become accustomed to the demands of their tasks. During this period they build up a level of stamina or 'task fitness' which matches their task requirements – this is known as 'work hardening'. Sometimes the workload can increase suddenly and the operator does not have a period in which to adjust. As a consequence, their level of task fitness does not match the increased demands and the operator may find that they become overloaded. Working in an environment where operators have to deal with a workload with peaks and troughs may result in them being more susceptible to injury. Work rates should be kept as consistent as possible and the workload should be spread out evenly across a shift.

Operators returning to work after an absence (eg holiday, sickness or maternity leave) also need a period of adjustment. During the absence their body will not retain the same level of task fitness and they therefore require a gradual build-up period to become reaccustomed to the work demands without overloading themselves. The same principle should be applied to new starters.

Sudden changes to the work routine or environment can have similar effects. The body will have become accustomed to working in a particular way and will have developed a level of fitness for that procedure. However, if the task or environment is abruptly changed – even though the underlying process may remain basically the same – the body is suddenly required to carry out a different task. As a consequence, there is a mismatch between the body's task fitness and the demands of the new procedure. Unless an acclimatisation period is allowed, the operator may become susceptible to injury.

Personal factors

Certain hobbies and activities and some medical conditions may predispose the operator to developing a disorder. However, it is not uncommon for the ULD sufferer to find it hard to accept that a factor outside the workplace is the source of their condition.

The playing of certain musical instruments involves movements and postures which are likely to enhance the development of a ULD. The type of instrument will influence the site of the disorder. For example, pianists frequently suffer symptoms of pain in their hands and wrists similar to those experienced by keyboard operators. Violinists typically complain of neck pain from having to clamp the violin under their chin while holding their head at an irregular angle. Trumpeters and trombonists sometimes complain of shoulder problems from gripping the instrument and supporting its weight away from the body.

Other hobbies such as knitting, crocheting, bell-ringing and playing computer games are also associated with the development of ULDs. Certain sports have also been closely associated. Racquet sports in particular – squash and tennis top the list – have been highlighted as a likely cause of ULDs. However, as is the case with work-related causes, sufferers need to be involved in these activities for extended periods of time on a regular basis for them to have a bearing on the development of a condition. Clearly, few people spend as much time indulging in their hobbies as they do in their work, and opportunities for rest during the practice of a hobby are greater. It is often the case that hobbies 'add insult to injury' as the operator uses their limbs more extensively than colleagues who may use their non-work time to rest.

A knock or blow to a vulnerable area such as the wrist may make it more susceptible to developing an injury. Certain conditions, such as pregnancy, the menopause and diabetes, have also been associated with ULDs. Although not necessarily indicated as a precipitating factor, psoriasis has been shown to be a common condition of ULD sufferers.

There is a higher incidence of ULDs among female than male workers. However, this is likely to be a result of the fact that in many instances women outnumber men carrying out highly repetitive operations.

Responding to ULDs

ULDs can be permanently disabling. They normally become progressively worse unless action is taken to identify the source of the problem and to deal with it. Obviously, the most effective action is one which prevents the onset of the condition in the first place. However, this does not assist those organisations which are already faced with operator injury and which need to take appropriate action in response.

In the first instance it is important that a clear diagnosis is made by someone with a thorough understanding of the condition and who will not be swayed by subjective reports regarding the working conditions or task demands. An accurate diagnosis is necessary so that the individual can be given the appropriate treatment and so that the organisation knows exactly what it is dealing with. In some cases the organisation can be satisfied from the outset that the condition is not work-related.

Treatment
It is common for a ULD sufferer to be signed off work for a period of time to rest. Alternatively, the employee may be advised to return to work on light duties. If this is the case, the organisation should

ensure that these duties do not incorporate activities which are likely to aggravate the condition further. When an operator returns to work after an absence they should be given an acclimatisation period in which to become used to the demands of the task again. They should be closely monitored during this phase. Of course, if factors within the workplace have been identified as being the cause of the injury, they should be addressed prior to the individual's return. Consideration should be given to whether other individuals are experiencing similar problems.

Sufferers may be prescribed painkillers and may be advised to attend physiotherapy sessions. In some cases the affected limb will be immobilised in a splint or plaster cast. Injections into the limb are also offered for certain conditions – these are painful for some time after the injection has been administered. Surgery is an option in more severe cases. Irrespective of the treatment, ULD sufferers need to be reminded that certain medical interventions will suppress their symptoms and make them think that they can return to work before they should. Individuals should be guided by their GP or consultant.

Reporting symptoms

As most ULDs are progressive, it is essential that the organisation responds appropriately to cases of injury as soon as possible. The longer the operator is exposed to the conditions which have been identified as harmful, the more severe their condition will become and the less likely it is that they will return to their original health. To facilitate the response process, organisations should have a facility in place whereby operators can report symptoms at the earliest opportunity. Operators should be advised of the symptoms they may experience as a result of the task they are carrying out.

Operators should feel comfortable about reporting symptoms. It is in the organisation's own interests to find out as soon as possible that the operator is experiencing difficulties. Without treatment and action on behalf of the employer, the operator's condition will probably worsen, eventually resulting in their absence. This will have implications for the remaining workforce – and ultimately the sufferer may decide to make a claim against their employer.

Assessing risks

Organisations will have in place a mechanism for assessing risk and this should be used as a means of identifying areas where ULDs may occur. This will enable the organisation to take steps to reduce the risk. When reviewing the workplace to identify the potential causes of ULDs, the organisation should consider the workstation design, task demands, organisational factors, environmental conditions, tool and equipment design, and the contribution made by the operator. It is only by looking at the workplace as a whole that an accurate profile of the likely causes of ULDs can be developed.

On a positive note, most ULDs are caused by relatively simple aspects of the workplace and therefore, in most instances, addressing problem areas should be fairly straightforward. It is unlikely that ULDs will be eliminated completely but it is possible to reduce them significantly. Employers should undertake systematic risk assessments of their activities to reduce the risk of onset. Consideration of the factors discussed here will go a long way towards ensuring that the workforce remains free of musculoskeletal problems – providing appropriate action is taken.

IN SUMMARY

- ULDs can affect any area of the upper limbs from the fingertips to the neck.
- The main causes of ULDs are repetition, poor posture, force and static work.
- Other contributory factors include exposure to vibration or reduced temperatures, lack of appropriate rest and/or rotation, overtime, poor training and changing work demands without an acclimatisation period.
- Personal factors such as hobbies and certain medical conditions may predispose an individual to develop a ULD.
- Appropriate action should be taken once an individual reports symptoms of a ULD to prevent any further deterioration in the condition – and its possible onset in others.
- The assessment process should be used as a means of identifying the potential for ULDs in the workplace and appropriate action should follow.

REFERENCES AND FURTHER READING

Buckle P and Devereux J, 1999, *Work-related neck and upper limb musculoskeletal disorders*, European Agency for Safety and Health at Work, Bilbao.

Health and Safety Executive, 1990, *Work-related upper limb disorders: a guide to prevention*, HSG60, HSE Books, Sudbury.

Health and Safety Executive, 1994, *A pain in your workplace?*, HSG121, HSE Books, Sudbury.

Health and Safety Executive, 1998, *Musculoskeletal disorders in supermarket cashiers*, HSE Books, Sudbury.

Kuorinka I and Forcier L (eds), 1995, *Work-related musculoskeletal disorders (WMSDs): a reference book for prevention*, Taylor & Francis, London.

Moon S D and Sauter S L (eds), 1996, *Beyond biomechanics: psychosocial aspects of musculoskeletal disorders in office work*, Taylor & Francis, London.

Pheasant S, 1991, *Ergonomics, work and health*, Macmillan Press, London.

Putz-Anderson V, 1988, *Cumulative trauma disorders: a manual for musculoskeletal diseases of the upper limbs*, Taylor & Francis, London.

Tindall A, 1993, *Tenosynovitis: a case of mistaken identity*, Iron Trades Insurance Company Limited, London.

WORKING WITH EXTERNAL ERGONOMISTS

Certain organisations may not wish to undertake the required ergonomics interventions without outside assistance. There are various reasons for deciding to use an external service provider. The organisation may:

- lack in-house expertise;
- be working to a tight deadline with all its existing resources fully deployed;
- believe that the ergonomist has experience of similar situations elsewhere;
- consider that a 'fresh pair of eyes' can make a significant contribution to the process, especially when existing members of the project team have been working on the same problem for some time;
- hope to learn something about ergonomics;
- decide it is worth trying the approach – when all else has failed;
- hope to learn something about its competitors;
- be forced to use the ergonomist by an insurer or factory inspector;
- be trying to 'protect' itself against litigation by using an external 'expert'; and
- consider that outside consultants are easy to use and cost-effective in the long term (for example, recruiting a full time staff member may not always be appropriate).

The organisation may also have 'political' reasons for wanting to use a consultant: an external ergonomist should not 'have an axe to grind'; should be objective; and may help to resolve departmental conflicts.

The following guidance should help to improve the likelihood of having a beneficial relationship with an external consultant.

Finding an independent ergonomist

The HSE's booklet on selecting a health and safety consultant (HSE 1994) should prove useful – finding an ergonomist is a broadly similar process. There are several ways that consultants with ergonomics expertise may be selected:

- Via a recommendation (the best way of establishing the track record and appropriate experience of an ergonomist).
- Via the Ergonomics Society (it maintains lists of appropriately qualified individuals and organisations).
- Via the Yellow Pages or other local business telephone directory under 'Ergonomics', 'Health and safety' or 'Management consultants'.
- Via specialist exhibitions or advertisements in appropriate periodicals.
- Via the Internet – more and more consultants now have websites.

Because ergonomics is such a broad discipline, it would be unrealistic to expect all ergonomists to have an in-depth knowledge of every aspect of the subject. Ergonomists tend to specialise, like any other professional. It is important to ensure that the selected ergonomist has the appropriate skills to deal with the organisation's particular needs.

The process of working with external ergonomists

STAGE ONE: defining the problem

The ergonomist should be used to help to define the problem or project. An unambiguous proposal with specific terms of reference should be produced and agreed. It should cover:

- what will be done;
- how it will be done;
- who will actually do it;
- when it will be done;
- what the deliverable will be;
- what it will cost; and
- whether there are any conflicts of interest.

STAGE TWO: watching and learning

Obviously, no two projects are the same in terms of what is required or the resources available. What the ergonomist will do at this stage will vary depending on whether the client simply wants a few guideline suggestions for the height of a conveyor or is commissioning work on a new factory to be designed from scratch. The following should be seen within the context of the project being undertaken.

A common starting point for the ergonomist, when undertaking a workplace-based health and safety consultancy project of moderate size, will be some sort of task analysis to establish 'the world as it is' (and not the in-house perception of the situation).

There are various ways of doing this but typically it will involve:

- Direct observation.
- Indirect observation using video and stills photography.
- Measurements relating to:
  - dimensions of workstations, machinery, tools and equipment
  - temperature, lighting and noise levels
  - weights, volumes and centres of gravity
  - frequencies and cycle times.
- Consultation with:
  - all levels of management and supervision
  - the workforce and workers' representatives
  - health and safety, occupational hygiene and occupational health professionals
  - scientists, engineers, technologists, architects and designers
  - specifiers, procurers and marketing, sales and accountancy staff.

Questionnaires may be used (if considered appropriate) to establish:

- general levels of satisfaction with the work and workplace;

- under-reporting;
- specific 'problem' tasks;
- workforce perceptions about management motivation;
- acceptability and ranking of certain solutions; and
- work patterns, breaks and so on.

Questionnaires are only of value if they are well designed and used in an atmosphere of trust.

The ergonomist will probably want access to a wide range of information. Credible ergonomists are working to a code of professional ethics and the organisation should not therefore be concerned about divulging commercially sensitive information (withholding information will only reduce the quality of their contribution). The ergonomist should be willing to sign a standard confidentiality agreement.

STAGE THREE: coming up with solutions

Most ergonomists will, following the information-gathering phase, expect to spend a reasonable amount of time formulating their solutions. The output from this stage could be a design, drawing, prototype or model but more often than not culminates in the production of a report and sometimes a presentation.

The main purpose of a report is to:
- establish the current or predicted state of play;
- facilitate change through reasoned argument;
- make reasonably practicable, prioritised recommendations; and
- suggest a strategy for how this change should be achieved.

The report is likely to make recommendations that focus on areas such as:
- the physical aspects of the work and the workplace;
- the organisational aspects of the work;
- training and information needs; and
- benchmarking – comparisons of the system or product against: other systems, organisations or products; Acts, regulations, guidance and standards; and previous ways of working or versions of the product.

STAGE FOUR: implementing the solutions

It is not uncommon for the ergonomist to be excluded from this stage. However, it is often the case that compromises are needed during the implementation phase and the ergonomist is usually well placed to consider the various trade-offs and make recommendations.

STAGE FIVE: monitoring

The ergonomist may be involved at this point in repeating the task analysis elements employed during stage two and then undertaking a comparison. Once the monitoring stage has been successfully completed, any required modifications should become apparent. Using this feedback loop, a final and satisfactory solution should be produced.

It is not unusual at this stage for an organisation to ask the ergonomist to 'validate' or 'sign off' the changes that have been implemented. A competent ergonomist will not be concerned by such a request as long as all recommendations have been implemented.

Before an ergonomist is commissioned

The organisation should be prepared to:
- trust the consultant with confidential information –
  - give details of production methods and problems
  - provide detailed medical and sickness information
  - outline costs, profits, targets, outputs and future products;
- respond positively to 'unwelcome problems';
- act on the report and recommendations (bearing in mind that the report is 'discoverable' and may be used in any subsequent claims to the detriment of the organisation if recommendations have not been implemented); and
- invest resources in implementing the recommendations.

This last point was made clearly by the Chairman of the Public Inquiry considering the Clapham Junction railway accident in 1988 when 33 people lost their lives: "A concern for safety which is sincerely held and repeatedly expressed but nevertheless is not carried through into action, is as much protection from danger as no concern at all."

Making the best use of the ergonomist

The organisation should:
- not have a hidden agenda;
- be truthful;
- not hide anything;
- provide all the information requested;
- disclose future plans;
- make sure that ready access is provided to all key personnel; and
- ensure that everyone in the organisation knows in advance that the ergonomists will be on site and try to reassure staff that it is in everyone's interests to co-operate.

Depending on the history and culture of the organisation, staff may well be suspicious of 'outsiders'. It is perfectly natural for members of the workforce to be concerned about being sacked or made redundant, medically retired, moved to lower paid work or separated from their friends. Managers may also have concerns about being sued, losing output or spending money.

Apart from allaying fears, the other major advantage of briefing staff is that the ergonomist can start to work without having to spend time on explanations.

The independent ergonomist works best when:
- the client has genuine needs and intentions;
- the brief is well defined;
- there is free access to information and individuals; and
- it is possible to work co-operatively with other groups of professionals in the organisation.

The main advantages of using an independent ergonomist are: knowledge, experience, objectivity, authority and credibility.

REFERENCE

Health and Safety Executive, 1994, *Selecting a health and safety consultancy*, IND(G)133(L), HSE Books, Sudbury.

INDEX

abduction	136
access	13
adduction	136
adjustability	27, 57
age	22, 30, 110, 126
air hoses	85
anthropometric data	46
anthropometrics	13, 39
assessments	8, 109, 150
automation	98
autonomous working groups	95
backache	11, 57, 111
back belts	132
back injuries	33, 106, 112
body size	13
bonus systems	97
breaks	72, 84, 96
cable management	61
carpal tunnel syndrome	138
cervical spondylosis	141
chairs	12, 28, 57, 67
– armrests	58
– backless	59
– backrests	28, 57, 67
– five star base	58
– lumbar support	28, 58, 67
change management	99
clearance	14, 83, 124
coding	23, 122
colour	24
consultant ergonomists	153
controls	22
– compatibility	23
– functions	23
design	11
desks	59
– drawers	60
– layout	63
– surface area	59
– undersurface	61
disabilities	17, 126
displays	24
display screen equipment	57

document holders	62
Dupuytren's Contracture	139
dynamic muscle work	84, 145
environmental conditions	24, 32, 68, 124, 146
epicondylitis	140
epilepsy	73
exercise	74, 147
expectations	23
feedback	22, 30, 100, 129, 156
fitness	22
– task	96, 148
footrests	12, 28, 43, 59, 62, 67
force	20, 22, 78, 81, 86, 119, 144, 146
frozen shoulder	140
ganglion	139
gender	22, 125
glare	33, 63, 65, 69, 74
handedness	17, 82
handle construction	80
handle design	79
handling aids	114
hearing impairment	32, 70
hernias	113
housekeeping	31, 113
humidity	32, 69
illuminance	33, 69
job design	91
job enlargement	95
job enrichment	95
keyboard work	60, 67, 71
– speed of operation	71
– style of operation	71
labels	24, 25, 118, 122
layout	22, 24, 31, 63, 119
lighting	33, 64, 68, 74, 125
luminance	33
lux	69

INDEX

maintenance	31, 68	standing	12, 41
manual dexterity	32, 85	static effort	12
manual handling	105	static muscle work	12, 84, 113, 145
maximum reaching distance	15	stature	13
mock-ups	30	stereotypes	23, 24
motivation	92, 97	stress	16, 23, 70, 74, 99
mouse mats	63, 72	supervision	31, 68, 132, 148
mouse use	71, 72	supination	136
multi-skilling	95	symbols	25, 122
noise	32, 70	task analysis	19, 22
normal distribution	40, 43	temperatures	28, 32, 61, 69, 81, 85, 124, 145
normal working area	16	tendinitis	140
optimum reaching distance	16	tenosynovitis	137
osteoarthritis	141	thermal environment	32, 70
overtime	97, 147	thoracic outlet syndrome	141
		tools, hand	77
pedals	14	– domestic	86
percentiles	40	– multi-functional	86
personal space	16	– off-the-shelf	86
piece-rate systems	97	torque response	86
posture	11, 21, 41, 67, 71, 80, 116, 123, 135, 143	training	22, 31, 68, 71, 86, 128, 148
		trials	30, 68
power grasp	77	triggers	83
precision grip	77		
pregnancy	73, 125, 138	ulnar deviation	136
primary work envelope	16, 22	upper limb disorders	11, 33, 57, 71, 135
pronation	136		
pulling	119	VDU arms	63
pushing	119	ventilation	70, 124
		vibration	33, 85, 146
radial deviation	136	viewing distance	19, 21, 60
reaching	13, 15, 117	visual acuity	21
repetition	94, 142	visual fatigue	68, 74
rotation	91, 94, 120, 143		
		wheelchair users	13, 17, 58
safety	29	working height	13, 20, 41, 59, 118
schedules	97	workload	96, 148
screen filters	63	work organisation	91, 120
seating	28, 42, 57	work rate	95
secondary work envelope	15	wrist extension	136
sharp edges	28, 61, 81, 83, 87, 121, 138	wrist flexion	136
shiftwork	98	wrist rests	63
sitting	12, 41		
skin irritations	73	zone of convenient reach	15